101 Topics in Current Chemistry
Fortschritte der Chemischen Forschung

Managing Editor: F. L. Boschke

QD1
.F74
v. 101
c. 1a

75-401099/s

chells

Host Guest Complex Chemistry II

Editor: F. Vögtle

With Contributions by
R. Hilgenfeld, R. M. Kellogg, D. Landini,
F. Montanari, G. R. Painter, B. C. Pressman,
F. Rolla, W. Saenger

With 140 Figures and 40 Tables

Springer-Verlag
Berlin Heidelberg New York 1982

This series presents critical reviews of the present position and future trends in modern chemical research. It is addressed to all research and industrial chemists who wish to keep abreast of advances in their subject.

As a rule, contributions are specially commissioned. The editors and publishers will, however, always be pleased to receive suggestions and supplementary information. Papers are accepted for "Topics in Current Chemistry" in English.

ISBN 3-540-11103-4 Springer-Verlag Berlin Heidelberg New York
ISBN 0-387-11103-4 Springer-Verlag New York Heidelberg Berlin

Library of Congress Cataloging in Publication Data. (Revised) Main entry under title: Host guest complex chemistry.
(Topics in current chemistry = Fortschritte der chemischen Forschung; 98, 101).
Includes bibliographical references and index.
1. Electron donor acceptor complexes — Addresses, essays, lectures.
I. Vögtle, F. (Fritz), 1939 — II. Blasius, Ewald — III. Series: Topics in current chemistry; 98, etc.
QD1.F58 540s [541.2'242] 81-8864 AACR2

This work is subject to copyright. All rights are reserved, whether the whole or part of the material is concerned, specifically those of translation, reprinting, re-use of illustrations, broadcasting, reproduction by photocopying machine or similar means, and storage in data banks. Under § 54 of the German Copyright Law where copies are made for other than private use, a fee is payable to "Verwertungsgesellschaft Wort", Munich.

© by Springer-Verlag Berlin Heidelberg 1982
Printed in GDR

The use of registered names, trademarks, etc. in this publication does not imply, even in the absence of a specific statement, that such names are exempt from the relevant protective laws and regulations and therefore free for general use.
2152/3020-543210

Managing Editor:

Dr. *Friedrich L. Boschke*
Springer-Verlag, Postfach 105280, D-6900 Heidelberg 1

Editorial Board:

Prof. Dr. *Michael J. S. Dewar*	Department of Chemistry, The University of Texas Austin, TX 78712, USA
Prof. Dr. *Jack D. Dunitz*	Laboratorium für Organische Chemie der Eidgenössischen Hochschule Universitätsstraße 6/8, CH-8006 Zürich
Prof. Dr. *Klaus Hafner*	Institut für Organische Chemie der TH Petersenstraße 15, D-6100 Darmstadt
Prof. Dr. *Edgar Heilbronner*	Physikalisch-Chemisches Institut der Universität Klingelbergstraße 80, CH-4000 Basel
Prof. Dr. *Shô Itô*	Department of Chemistry, Tohoku University, Sendai, Japan 980
Prof. Dr. *Jean-Marie Lehn*	Institut de Chimie, Université de Strasbourg, 1, rue Blaise Pascal, B. P. Z 296/R8, F-67008 Strasbourg-Cedex
Prof. Dr. *Kurt Niedenzu*	University of Kentucky, College of Arts and Sciences Department of Chemistry, Lexington, KY 40506, USA
Prof. Dr. *Kenneth N. Raymond*	Department of Chemistry, University of California, Berkeley, California 94720, USA
Prof. Dr. *Charles W. Rees*	Hofmann Professor of Organic Chemistry, Department of Chemistry, Imperial College of Science and Technology, South Kensington, London SW7 2AY, England
Prof. Dr. *Klaus Schäfer*	Institut für Physikalische Chemie der Universität Im Neuenheimer Feld 253, D-6900 Heidelberg 1
Prof. Dr. *Georg Wittig*	Institut für Organische Chemie der Universität Im Neuenheimer Feld 270, D-6900 Heidelberg 1

Foreword

This volume II "Host Guest Complex Chemistry" continues the research reviews on crown type host/guest interactions. The introducing chapter included in volume I also gives some explanations and cross links to this volume. Topics here also are ranging from structural results to phase transfer phenomena, PT-catalysis and stereoselective reactions. The closing remarks are meant to hold for volume I and II.

Bonn, July 1981   F. Vögtle

Table of Contents

Structural Chemistry of Natural and Synthetic Ionophores and their Complexes with Cations
R. Hilgenfeld, W. Saenger, Göttingen (FRG) 1

Dynamic Aspects of Ionophore Mediated Membrane Transport
G. R. Painter, B. C. Pressman, Miami (USA) 83

Bioorganic Modelling — Stereoselective Reactions with Chiral Neutral Ligand Complexes as Model Systems for Enzyme Catalysis
R. M. Kellogg, Groningen (Netherlands) 111

Phase-Transfer Catalyzed Reactions
F. Montanari, D. Landini, F. Rolla, Milano (Italy) 147

Closing Remarks 201

Author-Index Volume 101 203

Structural Chemistry of Natural and Synthetic Ionophores and their Complexes with Cations

Rolf Hilgenfeld and Wolfram Saenger

Abteilung Chemie, Max-Planck-Institut für experimentelle Medizin,
Hermann-Rein-Straße 3, D-3400 Göttingen, FRG

Table of Contents

1 Introduction: Ionophores as Host Molecules for Cationic Guests 3

2 Depsipeptides: The "Prime" Ionophores 5
   2.1 Valinomycin 5
   2.2 Enniatins and Beauvericin 12

3 Macrotetrolides: Cubic Cages for Alkali Ions 15

4 Polyether Antibiotics: Pseudo-Cavities 23
   4.1 Classification 23
   4.2 Monovalent Polyether Antibiotics 25
   4.3 Divalent Polyether Antibiotics 28

5 Crown-Ethers: Synthetic Macrocyclic Multidentates 38
   5.1 Basic Stereochemistry of the 1,4-Dioxa Group in Polyether Complexes 38
   5.2 The Ion-Cavity Concept 41
   5.3 Effects of Anion and Cation Type on Complex Structure 43
   5.4 Simultaneous Complexation of Metal Ion and Water or of two Metal Ions 46
   5.5 Sandwich Formation 48
   5.6 Ternary Crown Ether Complexes 49

6 Macropolycyclic Host Molecules: Cryptands and their Cation Complexes 49

7 Open-Chain Polyethers: Wrapping of Metal Ions 60
   7.1 Podands with Aromatic Donor End Groups 60
   7.2 Complexes of Podands Containing End Groups Capable
      of Hydrogen Bonding 65
   7.3 Linear Polyethers without Donor End Groups 69
   7.4 Tripode Ligands 71

8 Concluding Remarks 74

9 Acknowledgement 75

10 References 75

1 Introduction: Ionophores as Host Molecules for Cationic Guests

Although the antibiotics nigericin and lasalocid as first representatives of the naturally occurring ionophores were isolated as early as 1951 from *Streptomyces* cultures [1], it was only in the late sixties that the function of these membranes affecting compounds as complexing and transporting agents for alkali metal ions was established [2-4]. Once their outstanding properties as selective alkali cation carriers had been understood, however, they attracted the attention of numerous investigators, and many details have been learnt about their structures, complex formation and physiological activity. Furthermore, a considerable number of model compounds were synthesized, above all the so-called polyether ligands which did not only help to understand the mechanism of action of the naturally occurring ionophores but also led to the rapid development of phase-transfer techniques in organic chemistry, to mention only one application.

Ionophores can be characterized as receptors which form stable, lipophilic complexes with charged hydrophilic species such as Na^+, K^+, Ca^{2+} etc., and thus are able to transport them into lipophilic phases. for example across natural or artificial membranes. Very often, the processes of complexation and transport are highly specific: Many of the ionophores display the ability of discrimination between alkali metal ions of different size. Thus, the antibiotic valinomycin has a 10^4 times greater affinity to potassium than to sodium ions [5]. Both the specifity of complexation and the transport of highly polar entities into non-polar media requires certain structural features of the ligands, and in this review we shall endeavour to point out what these features are.

The elucidation of the spatial structures of ionophores and their ion complexes is essential for understanding the detailed mechanisms of their biological action. From this point of view, the nature of the conformational rearrangements accompanying complex formation is of special interest, because these very conformational changes contribute considerably to ion binding selectivity. For that reason, this review will focus on comparisons between ligand conformations in both the complexed and the uncomplexed state whenever data on both are available.

Since X-ray diffraction today is still the unique method of providing information of high accuracy on molecular conformations, we shall consider mainly results obtained from these studies. However, although in general representing a form of low energy, ligand conformations in crystals may differ from those in solution. Therefore, we shall try to compare both the solid state and the solution conformations if investigations of the latter have been carried out.

In view of the limited space available, this review cannot be encyclopaedic. Rather, we will select certain structures which we consider to be of interest. However, after each section we will give a compilation of published crystal structures to stimulate further studies.

Now, what are the essential requirements for ligand to be an effective ionophore?

Metal ion complexes of ionophores can be considered as host-guest complexes in which the guest entity is of spherical shape and entrapped in a cavity-like structure formed by the cyclic or open-chain host molecule. This cavity site can either be preformed to accept the metal ion without major conformational changes or it can adopt its final shape upon complexation of the cation, associated with structural rearrangements. In all cases, a mutual geometrical, topological fit between host and guest molecules is essential for adduct stabilization, the adduct being in general a 1:1 complex for the ionophores. For an alkali metal ion as a spherical guest, the optimum complementary structural feature is a cavity of corresponding size, lined with polar groups in order to provide maximum interaction through ion-dipole forces. The polar ligand groups, which usually contain electronegative atoms such as oxygen, nitrogen or, more rarely, sulfur, should be situated in such a way that they can step by step replace the solvation shell of the cation during complex formation. The exterior of the ligand molecules, however, should be lipophilic to provide an appropriate surface for the non-polar medium into which the metal ion is being transferred.

These structural features are maintained more or less by all the ionophores in their complexed forms. However, mention should also be made of exceptions to these general structural principles. Thus, in the case of some ionophores, cation complexes of other than 1:1 stoichiometry have also been described in which the formation of "sandwich"-type arrangements is favored [6].

Generally, two different modes of transmembraneous transport have been established: the "carrier" and the "channel" mechanism. The ionophores considered here act by the carrier mechanism. They form discrete antibiotic cation complexes at one interface of the membrane which then migrate across the membrane to the other interface where the metal ion is released. This kind of transport is displayed by the depsipeptide-type antibiotics which form positively charged complexes with metal ions. This is also true for the macrotetrolide nactins whereas the open-chain polyether antibiotics of the nigericin family mainly lead to electrically neutral metal ion complexes by dissociation of their carboxyl group. For the latter type of carriers, the ion transport of metal ions is coupled with a transfer of protons in the opposite direction.

Some linear peptides such as the gramicidins A, B, and C, alamethicin, suzukacillin, and trichotoxin A-40 do not act as carriers but they form transmembrane channels across which alkali metal ions can migrate. Just as the carrier cavities, these channels display a hydrophilic interior and a lipophilic exterior, but in contrast to the former they exhibit poor ion selectivity. Since no complete X-ray studies of any of these "channel" forming agents are available [7], only few facts are known about their conformations. Therefore, they will not be treated in this review.

A main section of this review will be devoted to the stereochemistry of synthetic ionophores and their ion complexes. The first analogs of the ionophorous antibiotics were cyclic *crown ethers* [12] some of which were shown to display similar selective complexation and transfer as the naturally occurring ligands [8–11]. Later, macrobicyclic oligoethers providing three-dimensional receptor cavities were designed [13]. Although they cannot be considered as model compounds for natural ionophores, these so-called *cryptands* are still very useful because, just as the crown ethers, they have found considerable application in organic chemistry as reagents influencing the rate and stereochemistry of reactions.

Finally, a number of open-chain polyether ligands — frequently called *podands* — has been synthesized [14]. From the viewpoint of a structural chemist, the latter are perhaps the most interesting synthetic ligands because of their high flexibility which often allows for an optimum adoption to the shape of the guest entity. Though some synthetic polyethers have been shown to transport alkali metal ions across lipid bilayers, the membrane-affecting properties of many of these ligands have not yet been investigated. This is why Ovchinnikov and coworkers prefer the name "complexones" rather than "ionophores" for these compounds in order to emphasize their alkali cation-chelating properties [15]. However, in accordance with the more general definition of an ionophore given above we shall use the latter notation for all ligands that are capable of transferring group IA and IIA metal ions to lipophilic phases (which do not necessarily have to be membranes).

Apart from metal ions, many of the ionophores are also capable of complexing neutral or charged organic guest molecules. In these complexes, interaction between host and guest is mainly achieved by hydrogen bonding and dipole-induced forces. In contrast to group IA and IIA metal ions, which show spherically symmetric electron-acceptor properties, binding sites are of a more directional character in the case of molecular guests.

Since adducts between organic molecules are discussed in other reviews [345, 346] they will not be treated here, i.e. the discussion is restricted to complexes between ionophores and metal ions.

2 Depsipeptides: The "Prime" Ionophores

2.1 Valinomycin

Valinomycin is a macrocyclic dodecadepsipeptide with 12 subunits — amino- and hydroxycarboxylic acids — which are connected by alternate peptide and ester bonds. It consists of three identical fragments D-HyIv-D-Val-L-Lac-L-Val^1 (Fig. 1). Although the valinomycin structure has been discussed in a recent review on complex formation of monovalent cations with biofunctional ligands [16], it is described here in some detail because the exceptionally high K^+/Na^+ discrimination displayed by the antibiotic is unequalled by any other ionophore which led to its designation as a "prime ionophore" [17]. Yet another reason is the fact that the conformations of valinomycin in various solvents and in lipid bilayer membranes are still the object of intensive research so that periodic reviewing is justified.

Preliminary crystallographic results on the potassium complex of valinomycin were first reported by Steinrauf et al. in 1969 [18], and in 1975, a more detailed study by Neupert-Laves and Dobler was published [19]. Both revealed close to threefold symmetry of the complex, the potassium ion being located at the center of the 36-membered ring (Fig. 2). The latter is folded into six β-turns which are stabilized by intramolecular N—H ... O=C hydrogen bonds of the common $4 \rightarrow 1$ type [20], i.e. reverse hydrogen bonding between the C=O of residue i and the N—H of residue i + 3. As can be seen from Fig. 5a, hydrogen bonds of this kind lead to the formation of 10-membered rings.

---

1 D-HyIv = D-hydroxyisovalerate

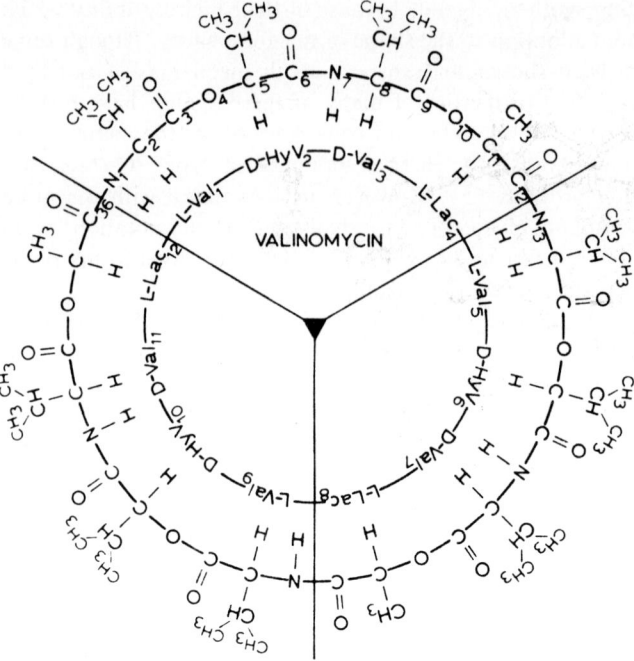

Fig. 1. Primary structure of valinomycin. The potential center of threefold symmetry is indicated

All the carbonyl oxygens involved in hydrogen bonding belong to amide groups; the ester CO groups are coordinated to the metal ion in an almost perfect octahedral arrangement, thus providing effective screening of the central cation from solvent interaction. All the lipophilic side chains are oriented toward the periphery of the molecule. Space filling models of the complex molecule are shown in Fig. 3.

The six hydrogen bonds are responsible for the limited flexibility of the depsipeptide ligand in the complex. Thus, valinomycin is not able to adjust itself to cations of different radius which explains its high selectivity. Whereas K^+ is of optimum size to fill the ligand cavity, Na^+ is too small to fully interact with all the ester oxygens and cannot be complexed without an energetically unfavorable breaking of intramolecular hydrogen bonds. On the other hand, Cs^+ as guest entity is too large to fit into the host binding site (for ionic radii see Table 1) and thus valinomycin complexes cesium ions less readily than K^+. Presumably, association with Cs^+ results in steric strain in the ligand backbone and/or considerable weakening of hydrogen bonding.

Only rubidium whose ionic radius is close to that of potassium is bound at least as effectively as the latter. An X-ray structural analysis of the valinomycin-RbAuCl$_4$ complex revealed essentially the same structure as in the K^+ complex, the octahedral cage of the oxygen atoms being by only 0.04 Å wider due to minor rotations of the ester bonds [21].

Numerous spectroscopic studies [22-28] and a conformational energy calculation [29] of the solution conformation of the valinomycin-K^+ complex are in good agreement

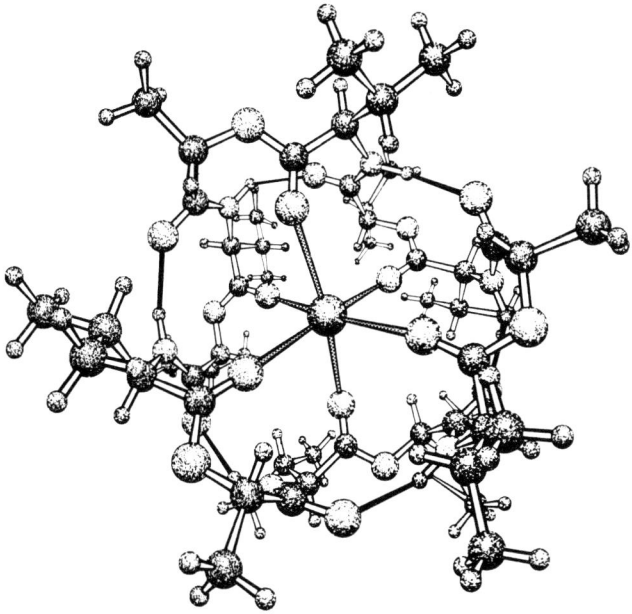

Fig. 2. Structure of the K^+ complex of valinomycin viewed from a 10 Å distance. Coordinative bonds are shaded, hydrogen bonding is indicated by narrow solid lines. Atomic radii are of arbitrary size and decrease in the order $O > N > C > H$. Similar graphical details are also followed in subsequent diagrams of molecular structures

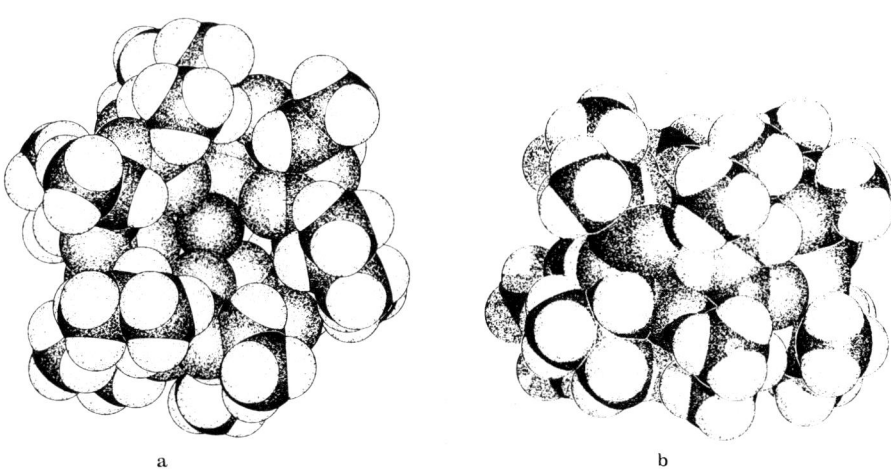

Fig. 3. Space-filling models of the valinomycin-K^+ complex: **a** view taken from the "top" of the molecule to demonstrate the optimum fit between host and guest, **b** view taken from one side of the complex showing the lipophilic periphery. Metal cation is completely covered and therefore not seen

Tab. 1. Ionic radii and charge densities of alkali and alkaline earth cations

Ion	r [Å]	Charge density × 10^{20} [Coulomb/Å³]
Li^+	0.68	12.1
Na^+	0.95	4.46
K^+	1.33	1.62
Rb^+	1.48	1.18
Cs^+	1.69	0.79
Be^{2+}	0.35	178
Mg^{2+}	0.66	26.8
Ca^{2+}	0.99	7.90
Sr^{2+}	1.12	5.46
Ba^{2+}	1.34	3.17

with the results provided by crystallographic methods. This is not true, however, for the uncomplexed form of the antibiotic.

As indicated in Fig. 1, there is a potential center of threefold symmetry in the primary structure of valinomycin. However, X-ray diffraction studies on the uncomplexed molecule [30-33] have shown that this symmetry is not maintained in the crystalline state. Instead, the macrocycle adopts a somewhat ellipsoidal shape with a pseudo-center of symmetry (see Fig. 4), which is stabilized by six intramolecular N—H ... O=C hydrogen bonds. Four of these belong to the above mentioned 4→1 type and involve amide NH and CO groups, the corresponding N ... O distances being in the range of 2.81–3.13 Å (mean 2.95 Å) [32]. The remaining two hydrogen bonds are of the rare 5→1 type, which was actually first discovered in valinomycin. Ester rather than amide carbonyl oxygens act as acceptors of these bonds which are considerably weaker than those of the 4→1 type. Their mean N ... O distance is 3.05 Å, and the N—H ... O angles deviate largely from linearity. In contrast to 4→1 type hydrogen bonds, those of the 5→1 type build up 13-membered rings in the structure, as can be seen from Fig. 5b.

These 5→1 bonds are largely responsible for the oval shape of uncomplexed valinomycin. Moreover, they direct two of the ester carbonyl oxygens toward the surface of the molecule (see Fig. 4). These might serve as initiators of complex formation by interaction with the metal ion prior to the placement of the latter in the ligand cavity. Based on this assumption, Smith and Duax developed a simplified model for the complexation process [33,34]. They proposed that the formation of the initial loose complex with the potassium ion is followed by cleavage of both the 5→1 type hydrogen bonds in order to enable all the other ester carbonyl oxygens to interact with the cation and to replace the molecules of its solvation shell one after another.

The three X-ray diffraction studies of uncomplexed valinomycin actually revealed structures of five independent molecules. They all exhibit largely the same conformation, although the crystals were grown from solvents of different polarity and showed different modes of molecular packing and solvent contents. This finding strongly suggests that crystal packing forces do not markedly affect the conformation of valinomycin molecules in the solid state and justifies the assumption that this

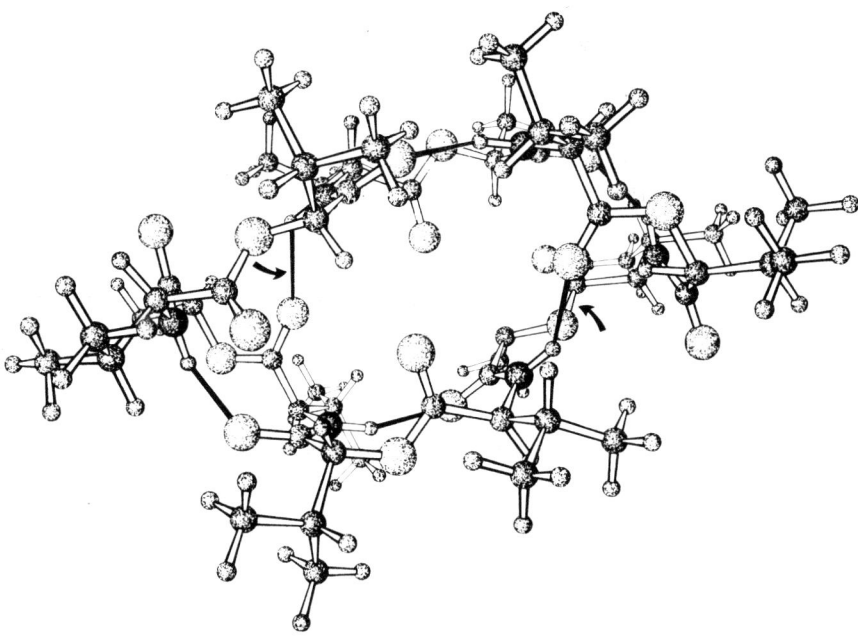

Fig. 4. Structure of uncomplexed valinomycin in the crystalline state. The two $5\to1$ type hydrogen bonds are indicated by arrows

low energy form also occurs in solution, at least in non-polar solutions where the solvent does not compete for hydrogen bonding [34].

There have been considerable efforts to elucidate the valinomycin conformation in various solvents by application of spectroscopic methods such as NMR, CD, ORD, IR, and Raman [35-45] as well as by conformational energy calculations [46]. However, little evidence has been provided for a solution conformation involving $5\to1$ type hydrogen bonds as observed in the solid state. Ovchinnikov and coworkers found that the conformation of uncomplexed valinomycin is highly dependent on solvent polarity. They proposed an equilibrium of three main forms which are usually referred to as A, B, and C (Fig. 6) [45]. Form A is believed to occur mainly in non-polar solvents such as CCl_4, octane etc: In the "bracelet"-like structure, there are six intramolecular NH ... O=C hydrogen bonds all of which are of the $4\to1$ type. Thus, in non-polar solvents the threefold symmetry of the molecule suggested by the primary structure (Fig. 1) is apparently maintained. However, Rothschild et al. reported evidence from Raman studies for hydrogen-bonded ester C=O groups in valinomycin dissolved in non-polar solvents [42] which indicates that the conformer found in the crystal is among those present in solution. Davies and Abu Khaled interpreted ^{1}H-NMR long-range proton coupling constants of valinomycin in $CDCl_3$ solution as consistent with a time-averaged conformation containing both $4\to1$ and $5\to1$ type H-bonds which are rapidly interconverting [35]. From conformational energy calculations, Maigret and Pullman concluded that the asymmetric conformer found in crystals is one of three possible structures existing in solution [46].

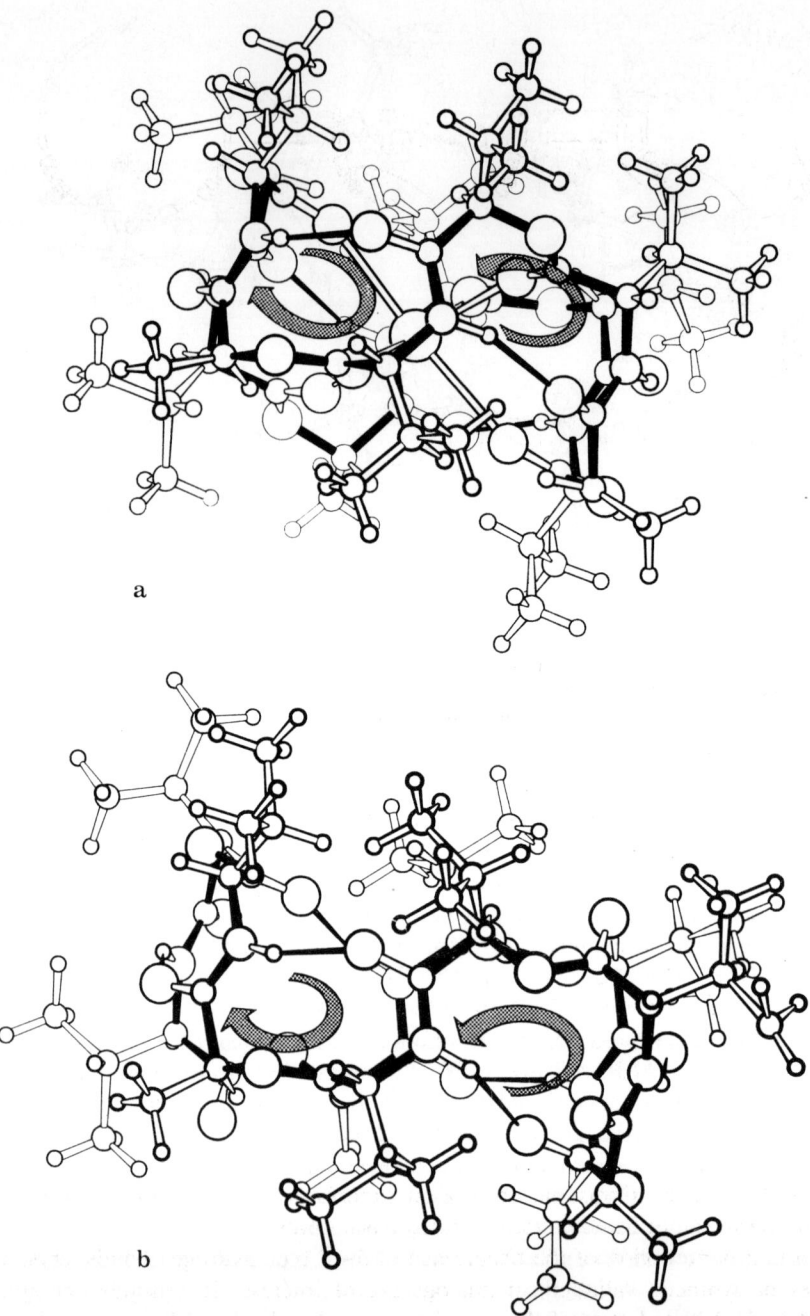

Fig. 5a and b. Hydrogen bonding in valinomycin-K^+ and in the free antibiotic: **a** Side view of the valinomycin potassium complex structure showing the ten-membered rings made up by $4\rightarrow1$ type hydrogen bonding. The antibiotic backbone is indicated by full bonds. **b** Side view of the uncomplexed valinomycin molecule. A 13-membered and a 10-membered ring, made up by $5\rightarrow1$ and $4\rightarrow1$ type hydrogen bonds, respectively, are shown

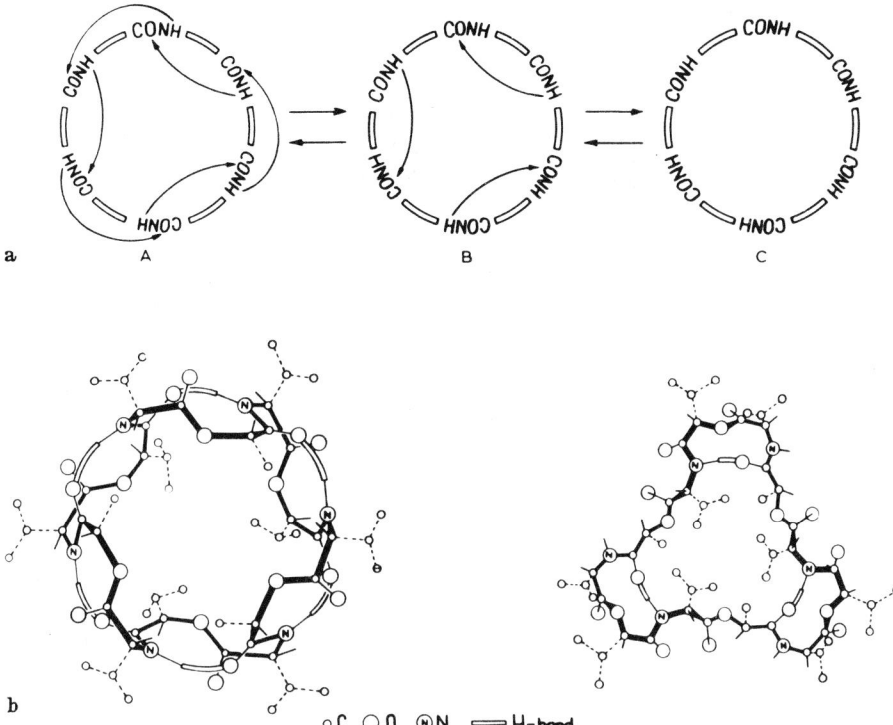

Fig. 6a. Equilibrium between the three basic solution forms of valinomycin (from Ref. [15]). **b** Form A (left) is believed to adopt a bracelet-like conformation, form B resembles a propeller (from Ref. [15])

Recently, ^{1}H-NMR spectroscopic results for uncomplexed valinomycin in a phospholipid bilayer revealed largely the same conformation as for the molecule in non-polar solvents. This suggests a preferential location of the free carrier in the apolar interior of the bilayer [43].

The "propeller"-like form B of the Ovchinnikov model predominates in solvents of medium polarity. There are only three intramolecular hydrogen bonds. Form C eventually was proposed for polar solutions where all the NH groups of the molecule are apparently hydrogen bonded to solvent molecules.

Besides these, more conformers exhibiting various degrees of hydrogen bonding have been detected by ultrasonic absorption experiments [44]; they are believed to represent intermediates between the described basic forms [23].

Concerning the question of the occurrence of $5 \rightarrow 1$ type hydrogen bonds, crystal structures of synthetic valinomycin analogs are of interest. Surprisingly enough, there are no hydrogen bonds of this type in uncomplexed meso-valinomycin where L-lactate is replaced by L-hydroxyisovalerate [47]. All of the six H-bonds are of the $4 \rightarrow 1$ type, and the molecule adopts a bracelet-like conformation. Only one $5 \rightarrow 1$ H-bond is reported for isoleucinomycin, in which the valinomycin Val groups are substituted by Ile [48]. Apparently, the latter model compound is an intermediate between the extreme conformations of valinomycin and meso-valinomycin in their crystalline states. This variety of conformations of highly similar compounds suggests that the energetic barriers between the different forms are in fact quite low. Evidence is added to this viewpoint by the observation of a second form of valinomycin crystals provided by Raman spectroscopic methods. This crystal form grown from o-dichlorobenzene shows no hydrogen bonding to ester carbonyl oxygens [49,50].

2.2 Enniatins and Beauvericin

Until recently, valinomycin has been regarded as a classic monocarrier which only forms complexes of 1:1 stoichiometry with alkali metal ions. However, in 1975 Ivanov reported evidence for the formation of adducts with a 2:1 valinomycin:cation ratio, and proposed a sandwich-type structure for the latter [6].

Somewhat better than for valinomycin, these sandwich complexes have been characterized in the case of the enniatin antibiotics the pecularities of which will be discussed in this section. In fact, though 1:1 enniatin:metal ion complexes have been shown to exist as well, it has been suggested that their membrane-affecting activity is due to the formation of sandwich aggregates [17].

The enniatins are cyclic hexadepsipeptides, i.e. 18-membered macrocycles, which exhibit considerable lower cation selectivity than does valinomycin. Thus, they are efficient ionophores for sodium as well as for potassium and caesium ions [51].

Four different species with closely related primary structures have been described: enniatins A, B, C and beauvericin. The latter is produced by the fungus *Beauveria bassiana* [52] whereas the former can be isolated from various *Fusarium* cultures [53]. Their general formula is cyclo-(L-MeX-D-HyIv)$_3$, where X is Ile for enniatin A, Val for enniatin B, Leu for enniatin C, and Phe for beauvericin (see Fig. 7). One important structural difference to valinomycin is the methylation of the amide

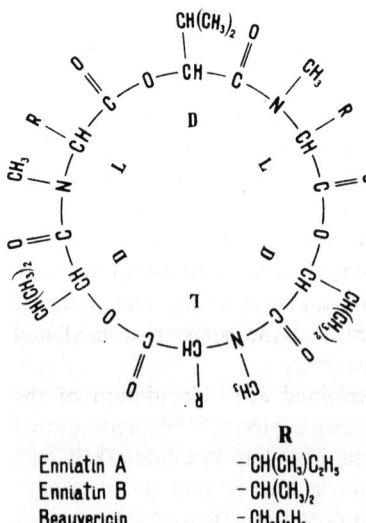

	R
Enniatin A	$-CH(CH_3)C_2H_5$
Enniatin B	$-CH(CH_3)_2$
Beauvericin	$-CH_2C_6H_5$

Fig. 7. Primary structure of enniatins and beauvericin

nitrogens. Thus, no formation of intramolecular hydrogen bonds is possible which makes the molecular backbone much more flexible.

According to Ovchinnikov and coworkers, it is this relative flexibility that accounts for the diminished selectivity of the enniatins toward alkali metal ions. Based on spectral data these authors postulated a structure of the 1:1 complexes with monovalent cations in which the metal ion occupies the center of the macrocyclic cavity and is coordinated by all of the six carbonyl oxygens in a octahedral arrangement [54,55]. Following this model, the ligand is able to easily adapt itself to the size of the metal ion by varying the orientation of the carbonyl groups. However, as pointed out by Steinrauf and Sabesan [21], this arrangement would lead to M^+ ... C (carbonyl) distances shorter than any observed earlier, especially in the case of the larger cations, but even for sodium. Moreover, the $C=O \ldots M^+$ angles would all be in the unfavorable range of 90–100°, as compared to 152–162° found for valinomycin-K^+.

The structure of the enniatin B complex with potassium iodide has been studied by X-ray crystallography [56]. Unfortunately, from this investigation it could not be concluded with certainty whether the metal ion is entrapped in the central cavity or, instead, occupies a site between two adjacent ligand molecules. An arrangement of the latter type has been observed in the crystal structure of the 1:1 complex between RbNCS and the synthetic LDLLDL isomer of enniatin B [57]. In this case, Rb^+ ions are coordinated by five carbonyl oxygens (three of the upper and two of the lower depsipeptide molecules) and the nitrogen atom of the isocyanate anion, thus forming infinite sandwiches.

Evidence is added to the assumption of a binding of the metal cation outside the macrocyclic cavity also in the 1:1 complexes by the fact that these, though being more stable than sandwich-type aggregates, are apparently ineffective in membrane transport [17]. This suggests that the metal ion is not sufficiently shielded from solvent or counter ion interactions which would be in agreement with a metal ion position outside the macrocyclic cavity.

In comparison to the 1:1 complexes, the enniatin sandwiches display a higher ion selectivity, their actual stability constants decreasing in the order $K^+ > Cs^+ > Na^+$. Besides adducts of 2:1 stoichiometry, a 3:2 "club sandwich" has been proposed for the Cs^+ complex [17].

Among the enniatin antibiotics, beauvericin is the one characterized in greater detail. This carrier is most interesting with respect to an *anion*-dependence of its transport properties [58]. Moreover, in contrast to valinomycin, it is capable of complexing alkaline earth as well as alkali metal ions [59]. A study of the effects of beauvericin on the conductivity of artificial lipid membranes in the presence of both mono- and divalent cations revealed a second-order relationship between conductance and antibiotic concentration [60,61]. Finally, Prince, Crofts, and Steinrauf detected an apparent charge of *plus one* for calcium in the beauvericin-mediated transport across bacterial chromatophore membranes [62].

These most unusual findings could be well explained after elucidation of the crystal structure of the beauvericin complex with barium picrate [61,63]. This adduct turned out to be a 2:2 dimer structure of the form $(Bv \cdot Ba \cdot Pic_3 \cdot Ba \cdot Bv)^+ Pic^-$ (Fig. 8) which is very unique inasmuch as three of the four picrate anions are incorporated into the space between the antibiotic molecules. Both the Ba^{2+} ions

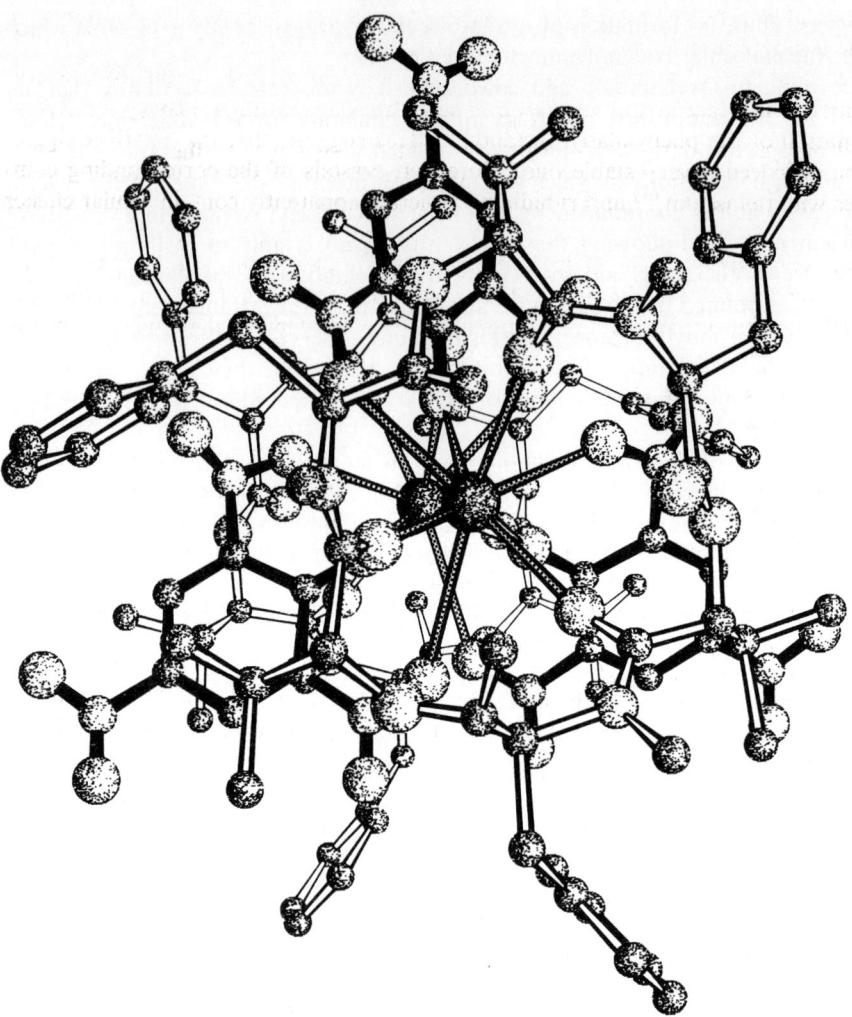

Fig. 8. Sandwich structure of the 2:2 complex between beauvericin and barium picrate shown from a 30 Å distance. From "top" to "bottom": 1) the "upper" beauvericin molecule (strong open lines), 2) one Ba^{2+} ion (dark sphere), 3) three picrate anions (full bonds), 4) the other Ba^{2+} ion, 5) the "lower" beauvericin molecule (weak open lines). For clarity, hydrogen atoms have been omitted

are *not* situated in the macrocyclic cavities but are displaced toward the center of the dimer, being connected to each other at a distance as short as 4.13 Å by the bridging picrate anions. Each of the metal ions is ninefold coordinated by the three amide oxygens of the ionophore (Ba ... O distances 2.64–2.77 Å), by three phenolate oxygens (one from each picrate anion, Ba ... O distances 2.72–2.78 Å), and by three nitro group oxygen atoms, again one contributed by each picrate (Ba ... O distances 2.96–3.07 Å).

The outer surface of the beauvericin-barium picrate complex is highly hydrophobic due to the orientation of the isopropyl, N-methyl, and phenyl groups toward the exterior of the molecule. The bulky phenyl residues provide an effective screening of the enclosed picrate ions.

A comparable 2:2 sandwich complex has also been found in a second crystal modification which showed the same overall structure only distinguished by a different orientation of the phenylalanyl residues [64]. This suggests that the described aggregation is indeed a very stable one. Moreover, crystals of the corresponding complexes with potassium [61] and rubidium [65] picrate apparently contain similar cluster units.

It is very likely that this beauvericin metal ion complex as revealed by X-ray structure analysis is the membrane-active species as well, because it is actually the only possible structure that is able to rationalize the aforementioned pecularities of ion transport induced by beauvericin. Above all, this is obvious for the residual net charge of plus one for the ionophore complex. *In vivo*, the picrate ions might be replaced by carboxylates which exhibit a similar chelating ability.

The structure of uncomplexed beauvericin as well has been investigated by single-crystal X-ray diffraction [65]. Apart from minor rotations of the side chains, its conformation is almost identical to that of the complexed antibiotic. Similar conclusions have been arrived at from a spectroscopic study of the free ligand in polar solvents [54] and from conformational energy calculations [65,66].

From the structural studies of beauvericin and its picrate complexes two things can be learnt:

1) The intriguing power of X-ray crystallography as a method of explaining the structure-activity relationships of biomolecules: It was only the X-ray structure of the barium picrate complex of beauvericin that provided an understanding of the anion-dependent activity of the antibiotic. All spectroscopic approaches failed to do so.

2) The different modes of binding displayed by the ionophores: Whereas valinomycin undergoes conformational rearrangements including the breaking of hydrogen bonds to replace the solvation shell of the metal ion and to finally enclose the latter in the macrocyclic cavity, beauvericin does not adjust its conformation to fit the ion but lipophilizes the polar guest entities by encapsulating the cation-anion complex as a whole. Since the host cavity does not maintain its function as selective binding site, this process is less cation specific. In view of the involved anion specifity, however, it still represents a high stage of molecular organization.

3 Macrotetrolides: Cubic Cages for Alkali Ions

In the cases of valinomycin and enniatin depsipeptides described in Section 2. the explanation for the structural origins of the more or less pronounced ion selectivities exhibited by these antibiotics was, though being plausible, somewhat tentative, as completed X-ray analyses were available only for complexes with a single ion species, i.e. with K^+ for valinomycin and Ba^{2+} for beauvericin. However, a detailed discussion of the structural features that lead to metal ion selectivities should be based on a whole set of comparable data on complex structures with various metal ions of

Table 2. Published X-ray structures of depsipeptide complexes: Number and lengths of coordinative bonds. Bond distances in this and subsequent tables were either taken from the original publication or calculated employing the Cambridge Crystallographic Data Package [343]. In cases where no numerical values are given, both these sources did not provide crystallographic coordinates. A dash indicates that there is no coordination to the heteroatoms in question. All distances are given in Å units

Compound	Coord. no.	Bonding distances			Comments, if any	Ref.
		M – O	M – anion	M – solvent		
[valinomycin · K]$^+$ I$_3^-$/I$_5^-$	6	2.69–2.83	—	—		19)
[valinomycin · K]$^+$ AuCl$_4^-$	6		—	—		18)
[valinomycin · Rb]$^+$ AuCl$_4^-$	6		—	—		21)
[prolinomycin · Rb]$^+$ pic$^-$	6		—	—	peptide analog of valinomycin	67)
[enniatin B · K]$^+$ I$^-$	6	2.6 –2.8	—	—		56)
[LDLLDL-enniatin B · RbNCS]	6	2.88–2.98	3.06	—	infinite "sandwich" spiral	57)
enniatin B · 6.25 H$_2$O · Na$^+$Ni^{2+}(NO$_3^-$)$_3$	6	—	—	—	both Na$^+$ and Ni$^+$ are not situated in the central cavity	68)
[(beauvericin)$_2$ · Ba$_2$ · (pic)$_3$]$^+$ pic$^-$	9	2.64–2.77	2.72–3.07		2:2 "sandwich" structure	61, 63)
[antamanide · Li · (MeCN)]$^+$ Br$^-$	5	2.04–2.23	—	2.07		69, 70)
[Phe^4, Val^6-antamanide · Na · (EtOH)]$^+$ Br$^-$	5	2.25–2.36	—	2.28		69)

pic = picrate

Table 3. Uncomplexed depsipeptides investigated by crystallographic methods

Compound	Constitution	Comments	Ref.
valinomycin	cyclo-(D-Hylv-D-Val-L-Lac-L-Val)$_3$	contains two 5→1 type hydrogen bonds	30–33)
isoleucinomycin	cyclo-(D-Hylv-D-Ile-L-Lac-L-Ile)$_3$	contains one 5→1 type hydrogen bond	48)
meso-valinomycin	cyclo-(D-Hylv-D-Val-L-Hylv-L-Val)$_3$	contains only 4→1 type hydrogen bonds	47)
(Me · Ala^2)valinomycin		conformation close to native valinomycin	71,73)
(Me · Ala^{2,6})octavalinomycin	cyclo-(D-Hylv-L-Ala-L-Lac-L-Val)$_2$	contains 5→1 and 3→1 type hydrogen bonds	72,73)
enniatin B	cyclo-(L-Me-Val-D-Hylv)$_3$	molecular cavity occupied by two water molecules	57,73–75)
DLLLLL-enniatin B			76)
beauvericin	cyclo-(L-N-MePhe-D-Hylv)$_3$	conformation very similar to that in Ba^{2+} complex	65)
antamanide	cyclo-(Val-Pro-Pro-Ala-Phe-Phe-Pro-Pro-Phe-Phe)	four water molecules serve as intramolecular bridges	77)
(Phe^4, Val^6)antamanide	cyclo-(Val-Pro-Pro-Phe-Phe)$_2$	three water molecules serve as intramolecular bridges	78–80)

Hylv = hydroxyisovalerate

different size. Fortunately, information of this kind has been provided for the macrotetrolide antibiotics by a large number of X-ray crystallographic studies of some of the free ionophores [81-83] as well as of their complexes with all the alkali metal ions (except lithium) [84-89] and with NH_4^+ [90-92].

The macrotetrolide antibiotics are 32-membered cyclic tetralactons which can be isolated from various *actinomyces* species. Five homologs of the general formula indicated in Fig. 9 are known, which are referred to as nonactin, monactin etc. depending on the number of methyl replaced by ethyl groups. The compounds are built up by four ω-hydroxycarboxylic acid subunits of alternating enantiomerism condensated to each other by esterification.

These ionophores which are frequently also called *nactins*, exhibit high selectivity in complex formation with alkali metal ions [93-94] as well as in ion transport through biological and artificial membranes [95,96]. In acetone, the stabilities of their K^+ complexes are increased by as much as 100 times compared to those of the corresponding Na^+ complexes. It is of interest that this selectivity is even enhanced by additional ethyl groups attached to the ligand backbone [96]. The general selectivity sequence for alkali metal ions displayed by the macrotetrolides is $K^+ > Rb^+ > Cs^+ \approx Na^+ \gg Li^+$, i.e. identical with that for valinomycin. However, even more stable than the K^+ complexes are those with NH_4^+ as guest ion which is not true for valinomycin.

The most extensive structural data available on macrotetrolides refer to *nonactin* and *tetranactin*. The structures of both the antibiotics in their uncomplexed states have been analyzed by X-ray crystallography [81,82]. Surprisingly enough, their conformations differ considerably from each other. Whereas nonactin displays a spatial form with S_4 symmetry as indicated in Fig. 10, the tetranactin molecule is of somewhat elongated shape (Fig. 11). On the basis of force-field calculations the latter conformer was found to be more stable by about 25 kJ · mol^{-1} than the former [87]. A recent X-ray study of uncomplexed *dinactin* revealed an asymmetric structure which can approximately be described as an intermediate between the nonactin- and tetranactin-type conformations. Actually, the dinactin conformer has the lowest conformational energy among the three [83].

Spectroscopic investigations of the corresponding solution conformations [97-99] showed that free macrotetrolides display appreciable rotational freedom of the macrocyclic skeleton. Most probably, all types of conformers found in the crystalline state are among those present in solution.

Upon complex formation, the ionophores undergo considerable conformational rearrangements, mainly resulting in a turning of the ester carbonyl oxygens toward the interior of the molecule. Just as in the case of valinomycin, but in contrast to beauvericin, the central ligand cavity serves as binding site in all the ion complexes

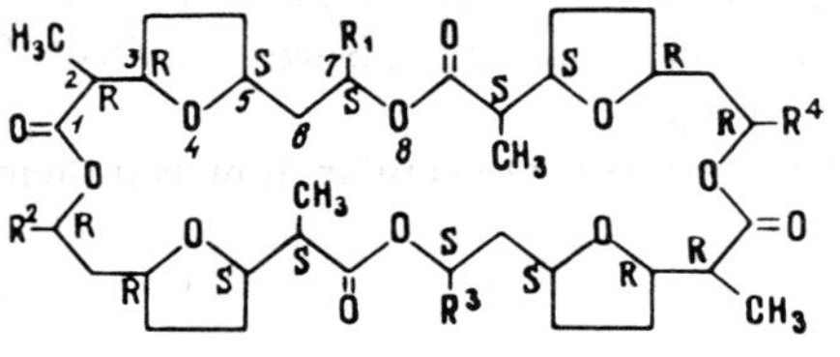

Nonactin: $R^1 = R^2 = R^3 = R^4 = CH_3$
Monactin: $R^1 = R^2 = R^3 = CH_3$, $R^4 = C_2H_5$
Dinactin: $R^1 = R^3 = CH_3$, $R^2 = R^4 = C_2H_5$
Trinactin: $R^1 = CH_3$, $R^2 = R^3 = R^4 = C_2H_5$
Tetranactin: $R^1 = R^2 = R^3 = R^4 = C_2H_5$

Fig. 9. Primary structure of macrotetrolide nactins

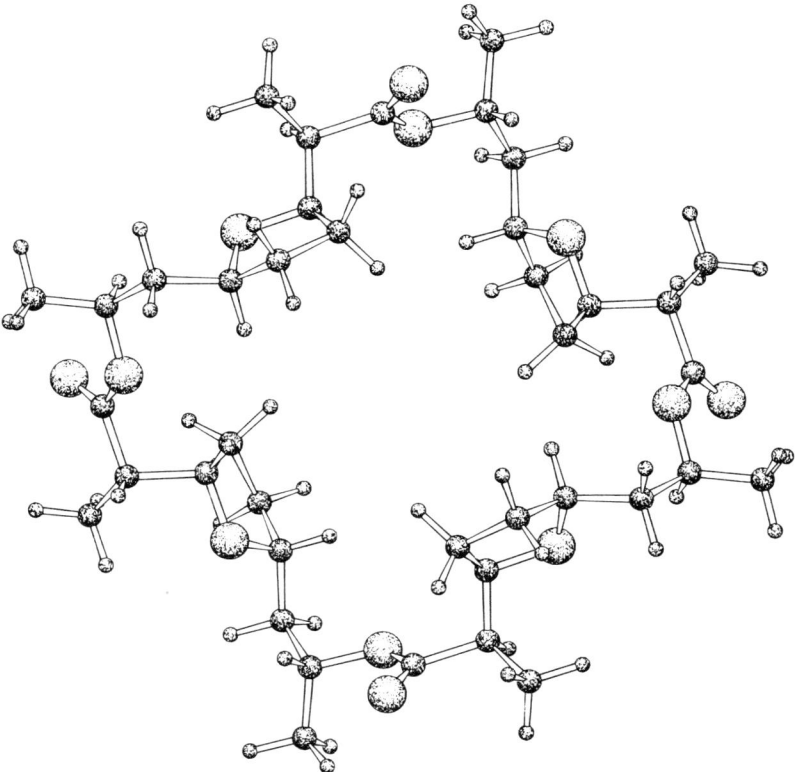

Fig. 10. Spatial structure of free nonactin

of the macrotetrolides. Though being distinct in the free state, nonactin and tetranactin exhibit very much the same conformation in their alkali cation complexes. Moreover, the overall structures of the complexes are retained upon replacement of one alkali ion by another. As an example, the spatial structure of the nonactin-Na^+ complex is shown in Fig. 12. The guest ions are eightfold coordinated in a more or less distorted cubic arrangement by the four ester carbonyl oxygens and by the four ether oxygens from the tetrahydrofuran moieties. The structures display all the features required for an effective alkali metal ion transport. The periphery of the complex molecules is highly hydrophobic. This holds above all for the tetranactin complexes where the ester oxygens are shielded from the exterior by ethyl groups [87] whereas nonactin lacking the latter presents a somewhat less lipophilic surface in its complexes which may account for the above mentioned diminished stability of nonactin complexes as compared to those of tetranactin.

Now, if the main spatial structures of all the complexes are similar, how can their different stabilities be rationalized? It is in fact the structural niceties that are responsible for the cation selectivity. In the K^+ complexes [84, 85, 87], the distances between the potassium ion and the ether oxygens are slightly larger than those to the carbonyl oxygens, the latter being close to the sum of the ionic radius of K^+

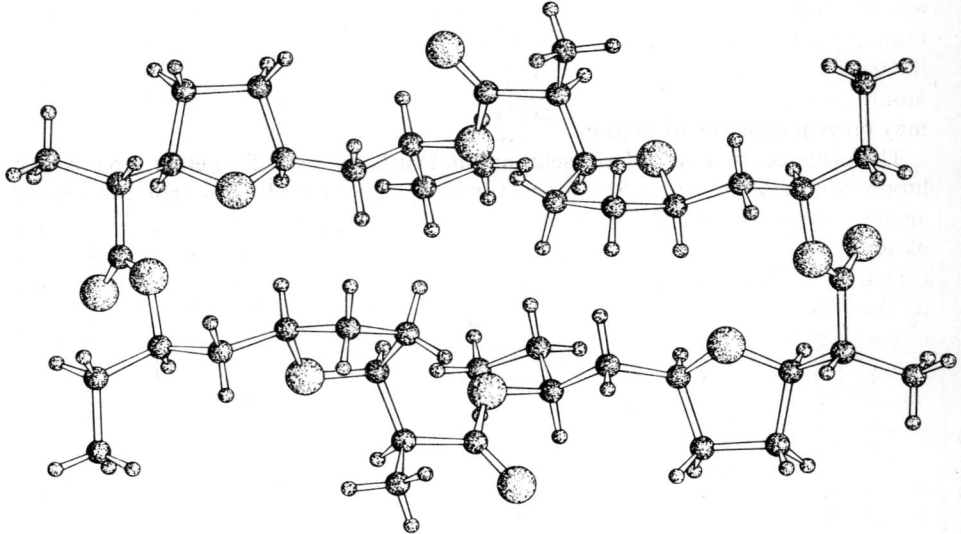

Fig. 11. Molecular structure of tetranactin in the crystalline state

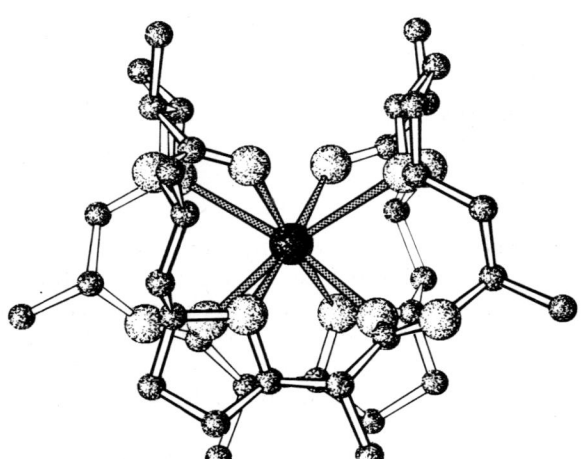

Fig. 12. Structure of the nonactin complex with Na^+. Hydrogen atoms are not shown

(1.33 Å) and the Van der Waals radius of oxygen (1.40 Å). When potassium is replaced by sodium, the carbonyl oxygen atoms approach the metal ion (Na^+ ... O distances 2.40–2.45 Å), but the ether oxygens are obviously restricted in doing so due to intramolecular steric hindrace [86, 87]. The Na^+ ... O (ether) distances are as long as 2.70–2.94 Å, i.e. very much larger than 2.35 Å, the sum of the Van der Waals radius of oxygen and ionic radius of sodium, thus providing considerably less electrostatic interaction and leading to distortion of the cubic coordination arrangement. The ligand cavity is only partially able to adjust to the smaller size of the guest ion which accounts for the reduced stability of the Na⁺ complexes. Binding of lithium ions by nactins has not been observed [89]. Clearly, the ionic radius of Li⁺ (0.60 Å) is too small to provide favorable interaction with all the ligand oxygen atoms. Furthermore, the high hydration energy of Li⁺ ($\Delta G_H^\circ = -503$ kJ · mol⁻¹) may prevent complex formation.

The distances between alkali metal ion and carbonyl as well as ether type oxygen atoms as found in various X-ray structures of nonactin and tetranactin are plotted against the ionic radii of the guests in Fig. 13a. Incorporation of larger ions such as Rb⁺ and Cs⁺ into the macrocyclic cavity leads to its expansion which involves essentially a slight outward displacement of the carbonyl oxygens. This is reflected in increasing rotation angles about the ester linkages, as indicated in Fig. 13b for nonactin complexes. The observation of Cs⁺ ... O (ether) distances shorter than the sum of the ionic radius of Cs⁺ and the Van der Waals radius of oxygen in the cesium-tetranactin complex [89] suggests the presence of steric strain in the ligand molecule which would explain the diminished stability of the complex.

The solution conformations of the nactin complexes with alkali cations have been investigated by NMR [97, 98, 100, 101], IR [102], and Raman [103, 104] spectroscopic methods. All these studies revealed structures almost identical with those found in the crystalline state. Whereas in the alkali cation complexes of the macrotetrolides the distances between metal ion and ether oxygens are generally longer than the corresponding distances to carbonyl oxygens (see Fig. 13), the opposite is true for the NH_4^+ complexes [90-92]. In the latter, the guest moiety is complexed by four strong N—H ... O hydrogen bonds to the *ether oxygens* (N ... O distances ≈ 2.86 Å), while it

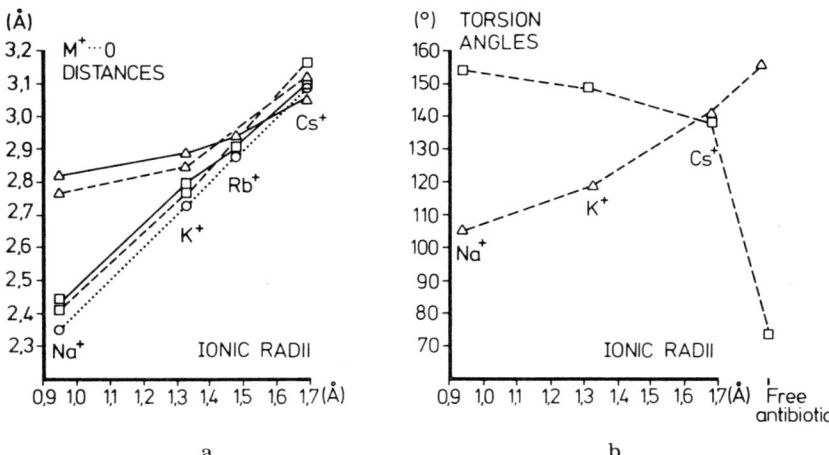

Fig. 13. **a** Distances between alkali metal ion and ligand donor atoms in macrotetrolide complexes as observed by X-ray crystallographic studies. Values for the nonactin (broken lines) and the tetranactin (solid lines) complexes are plotted against the radii of the cations. △, Mean distance between the cation and ether oxygen atoms. □, Mean distance between the cation and carbonyl oxygen atoms. ○, Sum of the ionic radius of the central cation and the oxygen Van der Waals radius (theoretical M⁺ ... O distance) (after ref. [89]). **b** Torsion angles about the ester linkages in nonactin complexes. Values for torsion angles about the C(7)–O(8) (△) and the C(1)–C(2) (□) bonds are plotted against the ionic radii of the cations. The atom numbering refers to Fig. 9

Table 4. Published crystal structures of nactin and of boromycin complexes

Compound	Coord. no.	Bonding distances		Comments, if any	Ref.
		M—O (carbonyl)	M—O (ether)		
[nonactin · Na]$^+$ NCS$^-$	8	2.40–2.44	2.74–2.79		86)
[nonactin · K]$^+$ NCS$^-$	8	2.73–2.81	2.81–2.88		84, 85)
[nonactin · Cs]$^+$ NCS$^-$	8	3.13–3.18	3.07–3.16		89)
[tetranactin · Na]$^+$ NCS$^-$	8	2.43–2.45	2.70–2.94		87, 88)
[tetranactin · K]$^+$ NSC$^-$	8	2.75–2.81	2.85–2.92		87, 88)
[tetranactin · Rb]$^+$ NCS$^-$	8	2.88–2.93	2.90–2.98		87, 88)
[tetranactin · Cs]$^+$ NCS$^-$	8	3.06–3.16	3.03–3.10		89)
[des-valino-boromycin · Rb]	8	—	2.80–3.17^a	*anionic* ligand is a Böeseken complex of boric acid with a macrodiolide	105, 106)

^{a} coordination distances to hydroxy and ester oxygens are included here

Table 5. Uncomplexed macrotetrolide antibiotics and boromycin

Compound	Comments	Ref.
nonactin	molecule shows S_4 symmetry	81)
dinactin	molecule adopts a twisted asymmetric conformation	83)
tetranactin	shape of the molecule fairly elongated	82)
des-valino-des-boron-boromycin	ligand backbone shows largely the same conformation as in Rb^+ complex of des-valino-boromycin	106)

interacts only weakly with the more distant carbonyl oxygens by $N^+ \ldots O^{\delta-}$ electrostatic forces (mean N ... O distances 3.00 Å–3.07 Å). Most probably, it is this additional host-guest hydrogen bonding interaction that accounts for the high stability of the NH_4^+-nactin complexes.

4 Polyether Antibiotics: Pseudo-Cavities

4.1 Classification

The ionophores we shall discuss in this section are referred to as *nigericin antibiotics* because nigericin was the first compound of this family to be discovered [1]. Other common designations are *carboxylic acid ionophores* and *polyether antibiotics*, describing essential structural features of these biomolecules which are unique among the ionophores described so far because they are linear and contain
1) a terminal carboxy group
2) one or two hydroxy groups at the other end of the molecule
3) several ether oxygen atoms provided by tetrahydrofuran and tetrahydropyran rings which may or may not be connected to each other by spiro-type junctions

In contrast to the above described neutral carriers which form positively charged complexes with cations, polyether antibiotics form neutral salts L^-M^+ with monobasic metal ions because their carboxy group is dissociated at physiological pH. Recently, however, it has been shown that they can also act as neutral ionophores to produce charged complexes such as LHM^+ [107,108].

Some of the carboxylic acid ionophores are capable of complexing divalent cations. Whether or not they have this ability can be used to classify the various members of the nigericin family. According to Westley [109,110], those ionophores *not* being able to transport divalent cations are referred to as *monovalent polyethers* some of which are shown in Fig. 14. These may be further divided into two subgroups, the distinction between which rests on whether the ionophore contains a hexapyranose moiety attached to the polycyclic ligand backbone or not. The former type antibiotics such as dianemycin, lenoremycin and A204A are called *monovalent monoglycoside polyether antibiotics*. The sugar unit is bound as an α-glycoside in antibiotic A204A, and as a β-glycoside in the other members of this class.

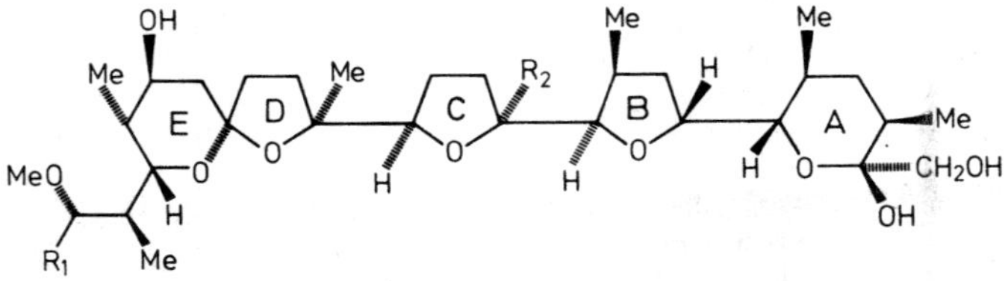

Monensin : $R_1 = CH(Me)CO_2H$; $R_2 = Et$

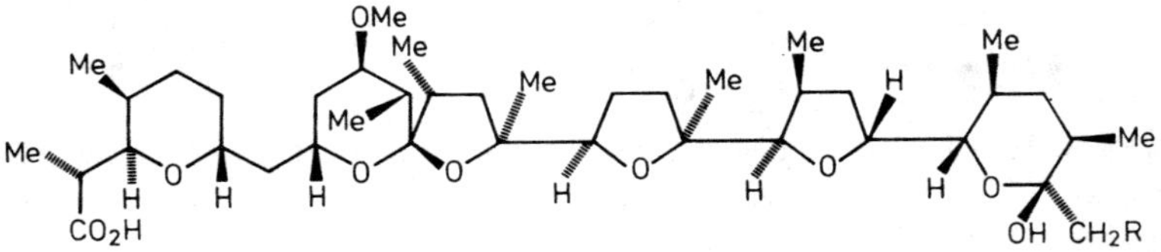

Nigericin : R = OH
Grisorixin : R = H

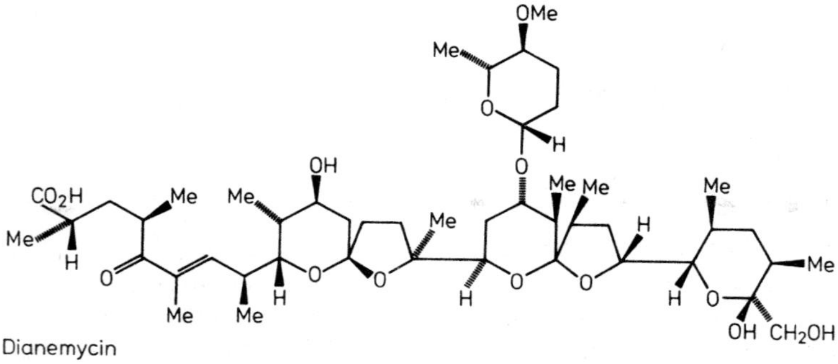

Dianemycin

Fig. 14. Some monovalent polyether antibiotics mentioned in the text

The *divalent polyether antibiotics* are much fewer in number than those of the first group. The most popular representatives are lasalocid, formerly known as antibiotic X-537A, and A23187. More recently discovered examples are lysocellin [111], ionomycin [112,113], and antibiotic X-14547 A [114] (see Fig. 15). Actually, the Ca^{2+}-specific ionophores A23 187 and X-14 547 A should be classified as pyrrole ethers rather than polyethers, because they possess only two and one ether functions, respectively. In fact, they are the only monocarboxylic bioionophores reported thus far to contain nitrogen.

The ion selectivities displayed by the polyether antibiotics are given in Table 6. They are somewhat lower than those of the nactins, but in some cases still considerable. Thus, monensin exhibits specifity for Na^+ ions whereas nigericin prefers K^+ [115,116].

The discovery of the calcium-transporting properties of lasalocid and A23 187 has prompted numerous studies concerning their physiological activities [117] and they were shown to be potential cardiovascular agents. Generally, polyether antibiotics are effective against gram-positive bacteria and fungi, and some of them have been claimed to be powerful insecticides or pesticides [109]. Although clinical applications

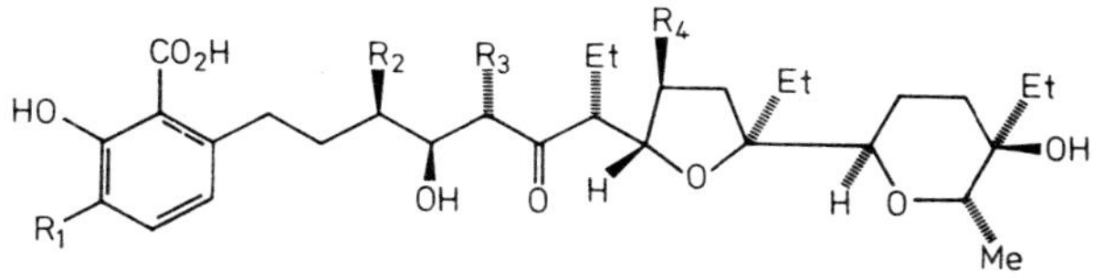

Lasalocid A : $R_1 = R_2 = R_3 = R_4 = Me$

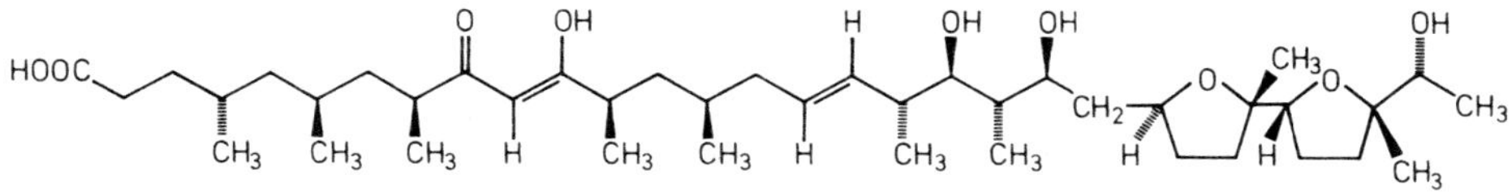

Ionomycin

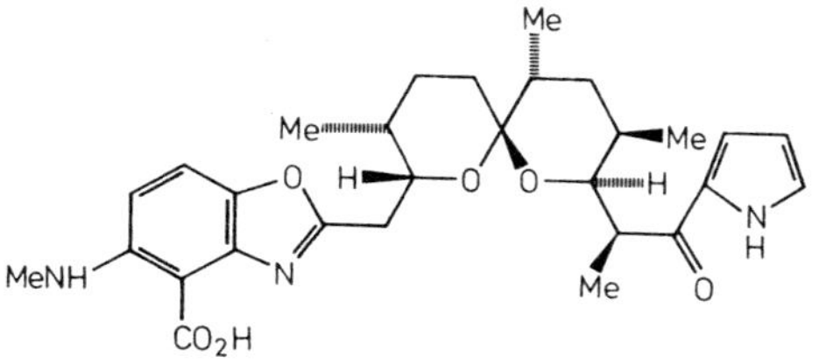

Antibiotic A23187

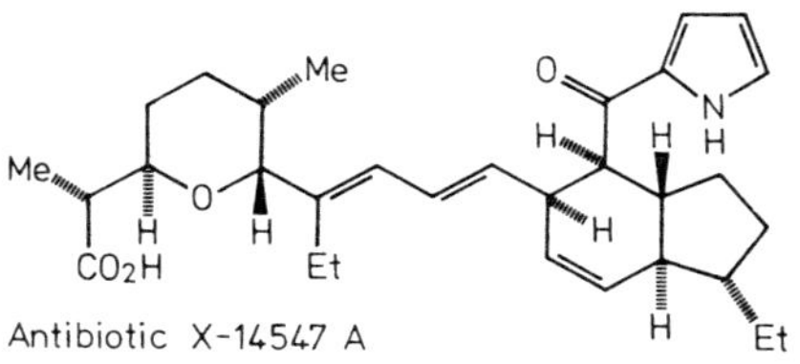

Antibiotic X-14547 A

Fig. 15. Divalent polyether and pyrrole ether antibiotics mentioned in the text

have been hampered by their parental toxicity, almost all of these antiotics have become particularly important as coccidiostatica for poultry industries.

It is probably these commercial applications that caused extensive search for new members of the nigericin family. Up to now, more than 45 distinct polyether antibiotics have been isolated from various *Streptomycetes*, and still a few ones are added every year. Fortunately, the structures of many of these have been elucidated by X-ray crystallography so that quite a large body of data on ligand conformation is available.

Table 6. Ion selectivity patterns found for polyether antibiotics

monensin	$Na^+ \gg K^+ > Rb^+ > Li^+ > Cs^+$
nigericin	$K^+ > Rb^+ > Na^+ > Cs^+ > Li^+$
dianemycin	$Na^+ \approx K^+ > Rb^+ \approx Cs^+ > Li^+$
lasalocid A	$Ba^{2+} \gg Cs^+ > Rb^+ \approx K^+ > Na^+ \approx Ca^{2+} \approx Mg^{2+} > Li^+$

4.2 Monovalent Polyether Antibiotics

Monensin was the first polyether antibiotic to have its structure solved by crystallographic methods. In the silver salt, the monensin anion is wrapped around the cation and held in this conformation by two strong *head-to-tail hydrogen bonds* between the carboxylate group and the two hydroxy functions of the terminal tetrahydropyran moiety as indicated in Fig. 16 [118,119]. Two water molecules are also involved in hydrogen bonding to the ligand. The silver ion is coordinated in an irregular arrangement by six oxygen atoms, four of which are of the ether and the remaining two of the hydroxy type. The carboxylate oxygens do not interact with the cation but are involved in hydrogen bonding with hydroxy groups at the other end of the molecule. The metal ion is thus entrapped in a hydrophilic cavity (to be more exact, in a pseudocavity owing to the linearity of the ligand), while the exterior of the molecule is highly lipophilic thus fulfilling the requirements for an effective transmembrane transport.

Recently, crystal structures of three different monensin complexes with sodium have been published [108,120]. One of these is a NaBr complex in which the carboxy group is not deprotonated, the antibiotic thus acting as a neutral ligand rather than as anion [108]. Apart from minor differences in the hydrogen bonding scheme, the molecular conformation turned out to be very close to that in the Ag^+ salt. This does also hold for an anhydrous and a hydrated form of the monensin sodium salt [120]. From these results, it appears difficult to deduce to which conformational properties the marked specifity for sodium is due. However, Steinrauf and Sabesan have carried out computer simulations based on the available crystallographic data [21]. They found that larger cations would lead to a rotation of the primary alcohol group of ring A (see Fig. 14 for notation) away from the guest ion, thus weakening the hydrogen bond to the carboxylate and exposing the hydrogen atom to the solvent. It seems conceivable that this could account for a diminished complex stability.

Evidence is added to this viewpoint by inspection of the structures of the *grisorixin* complexes with Ag^+ and Tl^+ (ionic radii: 1.26 and 1.47 Å, resp.) as revealed by X-ray analysis [121,122]. The cavity containing the cation is slightly dilated in the

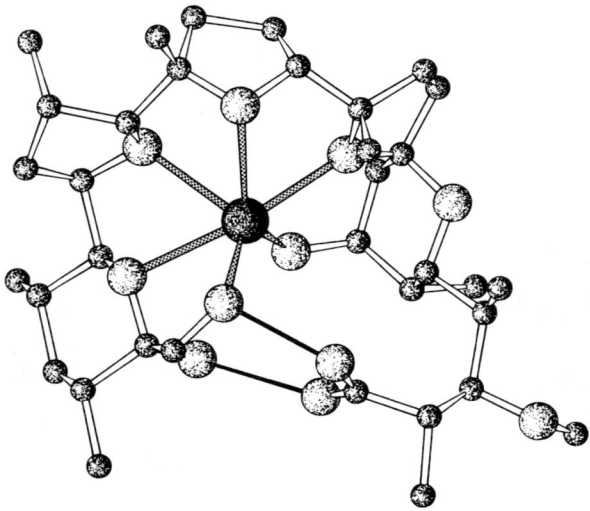

Fig. 16. Spatial structure of the Ag^+ complex of monensin. Note the two head-to-tail hydrogen bonds (full narrow lines) in the lower part of the figure. Two water molecules that are hydrogen-bonded to the terminal hydroxy groups have been omitted in the figure for the sake of clarity, as have been all hydrogen atoms

thallium salt, resulting in an orientation of one of the carboxylic oxygens toward the exterior of the molecule where it accepts a hydrogen bond of a water molecule not present in the Ag^+ salt. This indicates a somewhat less effective screening of the polar ligand groups from solvent interactions in the Tl^+ complex. Moreover, the length of the head-to-tail hydrogen bond is increased to 2.73 Å, as compared to 2.64 Å in the silver salt.

The free acid of monensin has also been subjected to an X-ray crystallographic study [123]. It revealed that the pseudo-cyclic conformation secured by head-to-tail hydrogen bonding is also present in the uncomplexed ionophore, the hydrogen bonding scheme, however, being somewhat different (see Fig. 17). One important feature of this structure is the presence of a water molecule located within the ligand cavity and hydrogen bonded to three oxygens of the antibiotic.

The conformation of the free ionophore is comparable to that found in the metal ion complexes. This is not surprising because of the relative rigidity of the ligand owing to the numerous methyl substituents attached to the ether rings. Moreover, the spiro junction between the D and E rings prohibits rotations. Therefore, this part of the molecule is in fact invariant in all the published crystal structures of monensin [120] whereas the rest of the ligand shows limited flexibility. Thus, even though changes of torsion angles are small, they add up to shift three oxygen atoms by as much as 2 Å in going from the uncomplexed to the complexed form.

Among the five cyclic ethers, ring C is the most flexible one as is indicated by the conformational changes it suffers upon complex formation. Based on this observation, Duax et al. proposed that the ion-entrapping process might be initiated by interaction of the ring C ether oxygen with the metal ion, followed by replacement of additional water molecules out of the hydration sphere by adjacent ether oxygens [120].

Spectroscopic studies including ^1H-, ^{13}C-, ^{23}Na-, ^7Li-, and ^{205}Tl-NMR [107, 124–127] have provided conformational assignments for monensin and its ion complexes in

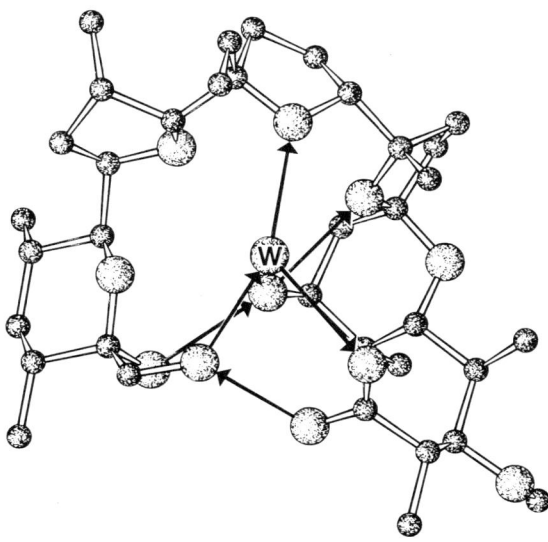

Fig. 17. Uncomplexed monensin shown in a similar orientation as for the Ag^+ salt in Fig. 16. A water molecule (indicated by "W") occupies the central cavity. Arrows mark the direction of hydrogen bonds from the donor to the acceptor atom. Hydrogen atoms are not shown

solution that are largely consistent with the results of X-ray work. Obviously, the free ionophore retains a water molecule also in chloroform solution, whereas no evidence was obtained for water being involved in the hydrogen bonding of the metal salts.

The above described structural properties of monensin and its metal complexes have essentially been found in all the monovalent polyether antibiotics. In some cases, e.g. nigericin and lonomycin, the carboxylate group is also involved in the coordination of the cation, thus providing additional ion-pair contribution to the binding energy. The head-to-tail hydrogen bonding which may include one or two hydrogen bonds is a common structural feature displayed by all the polyether antibiotics. Details of the coordination and hydrogen bonding patterns are given in Table 7a.

Monensin appears to be a good example to demonstrate how a biological host molecule utilizes a number of minor conformational changes in order to rearrange its binding sites for optimum complex formation with a guest entity.

This cooperative mechanism which may be compared with the induced-fit model of enzyme-substrate interaction [128] seems to be in some contrast to the nactins where only two torsion angles are subject to significant alteration (see Sect. 3).

4.3 Divalent Polyether Antibiotics

The term "divalent polyether" does *not* imply that these antibiotics are generally not able to complex monovalent cations, as can be seen from the ion selectivities given in Table 6. The most versatile antibiotics in the "divalent group" are the lasalocids which have been shown to effectively bind and transport Ba^{2+}, Ca^{2+}, Cs^+, Na^+ and Fe^{2+} [129] ions, and even catecholamines [130]. Interestingly, the barium ion, although the most strongly complexed, is not the most rapidly transported. This finding illustrates that in the ion-carrier mechanism tight metal ion binding of the ionophore ligand does not necessarily correlate with good carrier qualities. Rather it is of importance that not too strong binding allows for a sufficiently rapid release of the metal ion after having passed through the membrane — a subtle interplay of several components.

The lasalocids have several unique features such as a salicyclic acid moiety and a carbonyl group. Furthermore, their ligand backbones are considerably shorter than those of the other polyether antibiotics (see Fig. 15). This accounts for the inability of lasalocid ionophores to fully shield the complexed metal ion from the solvent by folding around it. The resulting complex has two distinct surfaces one of which is much more polar than the other, as demonstrated in Fig. 18 for lasalocid A, the most abundant species. In non-polar media, this unfavorable situation is overcome by the formation of dimeric complexes of stoichiometries $(M^+L^-)_2$ and $M^{2+}(L^-)_2$ for mono- and divalent metal ions, respectively. In fact, these aggregates present a highly lipophilic surface to the solvent.

An arrangement of this type was first shown by X-ray crystallography to exist in the barium complex of lasalocid A [131,132]. In this complex, two antibiotic anions coordinate to the Ba^{2+} ion in different ways. The metal ion occupies a polarophilic "pocket" mainly provided by one of the ligands and is coordinated by two ether, two hydroxyl and one of the carboxylate oxygens. The ninefold coordination is

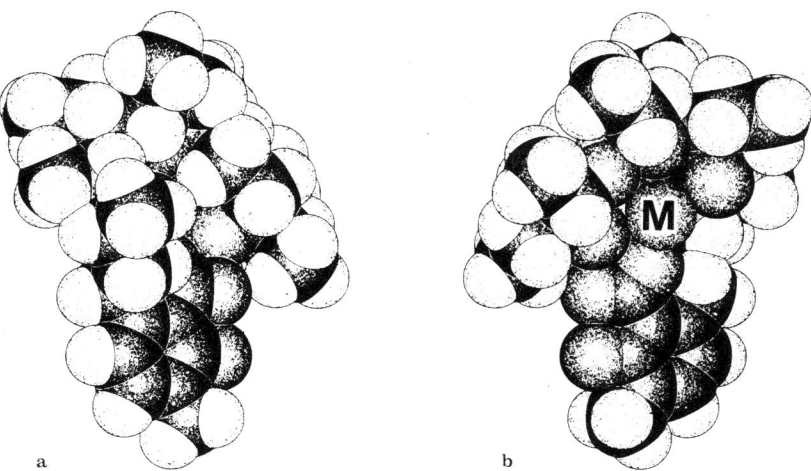

Fig. 18a and b. Space filling model of a 1:1 lasalocid complex with a metal ion (M): **a** from the non-polar side (metal ion completely covered) **b** from the polar side

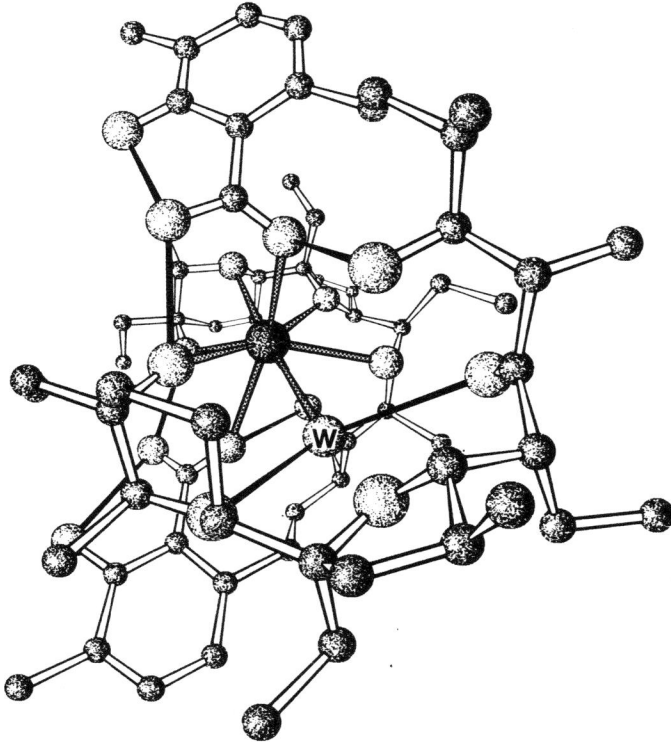

Fig. 19. Spatial structure of the Ba^{2+} complex of lasalocid A viewed from a 10 Å distance. The Ba^{2+} ion is mainly coordinated by the "far" ligand molecule to yield a 1:1 complex of the type shown in Fig. 18. This is covered by a second lasalocid molecule (the "near" one). The water molecule is indicated by a "W". Hydrogen atoms have been omitted for clarity

completed by two oxygens contributed by the second lasalocid anion, and a hydrogen-bonded water molecule occupying the cavity of the latter (see Fig. 19).

If the salicylate and the tetrahydropyran moieties are referred to as the "head" and the "tail" of the molecule, respectively, the two anions are arranged as a *head-to-tail dimer*. There are apparently no hydrogen bonds between the ligands, the latter being held together by their guest entities, i.e. the Ba^{2+} ion and the H_2O molecule. Each of the lasalocid molecules, however, is folded into a pseudo-cyclic conformation stabilized by an intramolecular hydrogen bond between the carboxylate and the terminal hydroxy group. This interaction is similar to the aforementioned "head-to-tail" hydrogen bond found in monensin and all the other polyether antibiotics belonging to this family.

Discrete dimers of the "head-to-tail" type have also been found in the crystal structures of the 2:2 lasalocid complexes with Ag^+ [133] and Na^+ [134] when crystallized from the nonpolar carbon tetrachloride. In both complexes, each metal ion is fivefold coordinated in the cavity of one anion, and there is an additional interaction with the second one, i.e. with its aromatic ring in the case of Ag^+ or with one of its carboxylate oxygens if the cation is Na^+.

However, when the sodium complex was crystallized from a solvent of medium polarity such as acetone, a *head-to-head dimer* structure was obtained [134]. This is also true for the free acid of 5-bromolasalocid [135] crystallized from C_6H_{14}/CH_2Cl_2 in which the antibiotic molecules are connected to each other by a hydrogen-bonded water molecule and by an additional hydrogen bond from the carboxy group of one ligand to the carbonyl group of the other.

It should be emphasized that in eighteen crystallographically independent molecular structures of lasalocid itself and of its cation complexes including the p-bromophenethylamine salt [130], the ionophore displays essentially the same conformation. Complexation of lasalocid by metal ions of different sizes is associated only with minor conformational changes of the ligand but to a much greater extent with configurational adjustment by changing the separation and orientation of the two ligand molecules relative to each other according to the spatial requirements of the guest cation. This simple mechanism provides proper host-guest adaptation similar, in a sense, to the complexation mode exhibited by the sandwich-forming beauvericin described in Section 2.2, and accounts for the versatility of lasalocid in complexing cations as different in size and charge as Na^+ and Ba^{2+}, or Cs^+ and Ca^{2+}.

The dimeric structures found in the crystalline state have also been detected in non-polar solutions by NMR spectroscopy [136] but the spectral data obtained in polar solvents where not consistent with such adducts [137]. Instead, they suggested the presence of simple 1:1 complexes. To assess the influence of solvent polarity, Paul and coworkers examined the crystal structures of free lasalocid and of its sodium complex crystallized from methanol, and, indeed, found these to be *monomers* [138, 139]. Both structures display the familiar head-to-tail hydrogen bonding. In the sodium complex, the Na^+ is coordinated to the same five ligand oxygens as in the dimeric structure, but it is capped by the oxygen of a methanol molecule.

Based on these structural data, it is reasonable to assume that monomeric forms are involved in cation uptake and release in polar media, e.g. at the exterior of the membrane, whereas the actual transport process is achieved by the lipophilic dimer [138]. A model for monomer — dimer transition has been provided by a recent

X-ray structural analysis of a 2:2:2 adduct between lasalocid, Na$^+$, and water obtained from 95% ethanol solution [140]. In this complex, each of the sodium ions is mainly associated with a single lasalocid anion in a manner reminiscent of the aforementioned monomeric Na$^+$ complex. Coordination of the metal ions is completed by two water molecules which are accommodated between the antibiotic molecules. The dimer is held together by hydrogen bonds *via* these water molecules. This suggests that the adduct might be a result of the initial association of two Na$^+$-lasalocid monomers, thus representing an intermediate in the complexation process.

The detection of both monomeric and dimeric structures in the solid state, depending on the polarity of the solvent from which the crystals were grown, and the consistency of these results with the solution structures as provided by spectroscopic methods nicely demonstrates that crystallographic data can be sensitive even to solvent influences. Moreover, the crystal structure of the "intermediate complex" shows the importance of solid-state conformational analysis for the elucidation of complex formation mechanisms or, more generally speaking, for molecular dynamics.

In contrast to lasalocid, antibiotic A23187 (see Fig. 15 for formula) does undergo considerable conformational changes upon transition from the free to the complexed state which is expressed in drastic alterations of four torsion angles [34]. The X-ray crystallographic analysis of the free acid [141] revealed a monomeric structure with the familiar head-to-tail hydrogen bond between the pyrrole nitrogen and the carboxy group.

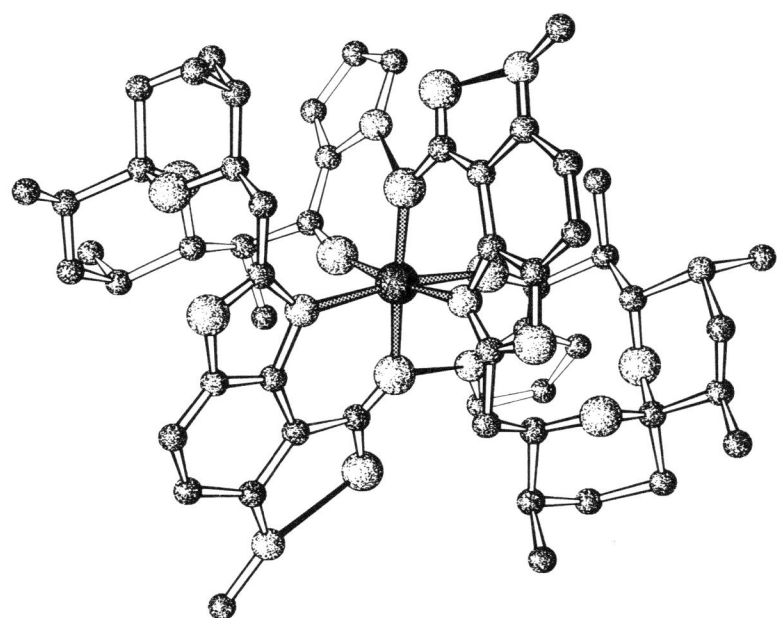

Fig. 20. The 2:1 complex between antibiotic A23187 and Ca^{2+}. Hydrogens are not shown

Table 7a. Coordination and hydrogen bonding in metal ion complexes of monovalent polyether antibiotics

Antibiotic	Cation	Total coord. number	Liganding oxygen atoms ether	hydroxy	carboxylate	Solvent	Number and lengths of "head-to tail" H-bonds	Comments, if any	Ref.
monensin	Na$^+$	6	4 2.36–2.54	2 2.34, 2.45	—	—	2 2.51, 2.64	refers to a hydrated sodium salt	120)
	Na$^+$	6	4 2.41–2.53	2 2.35, 2.38	—	—	2 2.58, 2.62	refers to an anhydrous form of Na$^+$ salt	120)
	Na$^+$	6	4 2.35–2.50	2 2.37, 2.42	—	—	2 2.73, 2.76	refers to NaBr complex	108)
	Ag$^+$	6	4 2.40–2.69	2 2.43, 2.45	—	—	2 2.51, 2.65	structure contains two H$_2$O molecules	118,119)
nigericin	Na$^+$	5	4 2.38–2.52	—	1 2.25	—	2 2.55, 2.75		146)
	Ag$^+$	5	4 2.47–2.66	—	1 2.26	—	1 2.59		147–149)
	K$^+$	7	5 2.67–3.09	1 3.06	1 2.58	—	2 2.63, 2.73	two of the oxygens interact only weakly with K$^+$	150)
grisorixin	Ag$^+$	5	4 2.4–2.7	—	1 2.20	—	1 2.64		121)
	Tl$^+$	5	4 2.6–3.0	—	1 a)	—	1 2.73	a water molecule is hydrogen-bonded to carboxy group	122)
lonomycin (emericid, DE-3936)	Na$^+$	6	4 2.40–2.51	—	2 2.38, 2.45	—	1 2.66		151)
	Ag$^+$	6	4 2.50–2.77	—	2 2.41, 2.65	—	1 2.73		151,152)
	Tl$^+$	6	4	—	2	—	1		153,154)
X206	Ag$^+$	6	2	1	—	—	1 2.69		155,156)

Ionophore	Cation	CN	n	d	n	d	n	d	n	d	n	d	Remarks	Ref.
alborixin	K⁺	8	3	2.76–3.07	4	2.69–2.98	1	2.89	—	—	1	2.64		157)
dianemycin	Na⁺	7	4		2		—		—		—			158)
	K⁺	7	4		2		—		—		—			158)
	Tl⁺	7	4		2		—		1 (H₂O)	—	—		A water molecule is inserted into the "head-to-tail" hydrogen bonding	158)
lenoremycin (Ro21-6150, A130A)	Ag⁺	8	6	2.46–2.88	2	2.38, 3.01	—		—		2	2.60, 2.67	pyranose oxygen is coordinated to the cation	159, 160)
A204A	Na⁺	6	4	2.72–2.85	—		2	2.71, 2.97	—		1	2.69	sugar ring well removed from the central cavity	161)
carriomycin	Tl⁺	6	4	2.82–3.00	—		2	2.73, 3.00	—		1	2.76	sugar ring does not interact with Tl⁺	162, 163)
K-41	Na⁺	6	4		—		2		—		1		the p-iodo- and p-bromobenzoate derivatives have been employed in the X-ray analysis	164)
6016	Tl⁺	6	4		—		2		—		1		glycoside moiety does not interact with the cation	165)

Table 7b. Divalent polyether antibiotics: Bonding distances in their complexes with metal ions. In the case of 2:1 complexes, the ionophore molecules are referred to as "A" and "B"

Antibiotic	Cation	Total coord. number		Liganding atoms and bonding distances (Å)					Solvent	Number and lengths of intra-ligand "head-to-tail" H-bonds	Comments	Ref.
				ether	hydroxy	car-boxy-late	keto	benzoxazole nitrogen				
lasalocid	Na$^+$	6	ligand A:	2	2	—	1	a)	—	1	refers to Na$^+$ salt crystallized from non-polar solvents; Structure is a 2:2 "head-to-tail" dimer	134)
			ligand B:	—	—	1	—	a)	—	1		
	Na$^+$	6	ligand A:	2	2	—	1	a)	—	1	refers to Na$^+$ salt crystallized from solvents of medium polarity. Structure is a 2:2 "head-to-head" dimer	134)
			ligand B:	—	—	—	1	a)	—	1		
	Na$^+$	6		2	2	—	1	a)	1 (MeOH)	1	refers to Na$^+$ salt crystallized from polar solvents. The structure is monomeric	138)
	Na$^+$ (1)	7	ligand A:	2 2.42, 2.47	2 2.56, 2.67	—	1 2.67	a)	1 (H$_2$O) 2.45	?	2:2:2 Na$^+$-lasalocid-water complex, cryst. from 95% ethanol. The two Na$^+$ ions show different coordination	140)
			ligand B:	—	—	—	1 2.41	a)	—	—		
	Na$^+$ (2)	6	ligand A:	—	—	—	—	a)	1 (H$_2$O) 2.37	?		

Structural Chemistry of Natural and Synthetic Ionophores and their Complexes with Cations

Ionophore	Cation	CN	Ligand						H₂O		Notes	Ref.
lysocellin	Ag⁺	5	ligand B: 2 2.40, 2.44	2 2.47, 2.72	—	—	—	a)	1 (H₂O) 2.40	—	2:2 "head-to-tail" dimer	133)
			ligand A: 2	2	—	—	—	a)	—	1 2.60		
			ligand B: —	—	—	—	—	a)	—	1 2.64		
	Ba²⁺	9	ligand A: 2 2.86, 2.98	2 2.71, 3.08	1 2.81	1 2.80	—	a)	—	1 2.72	2:1 (L:M) "head-to-tail" structure	131, 132)
			ligand B: —	1 2.84	1 2.64	—	—	a)	1 (H₂O) 2.74	1 2.77		
lysocellin	Ag⁺	6		2 2.38, 2.46	2 2.45, 2.55	1 2.55	1 2.99	—	a)	—	1 2.62	1:1 complex	166, 167)
ionomycin	Ca²⁺	6		1 2.45	2 2.44, 2.44	1 2.28	2 2.26, 2.28	a)	—	1 2.66	intermolecular hydrogen bonding leads to 2:2 dimers	145)
	Cd²⁺	6		1 2.38	2 2.38, 2.40	1 2.30	2 2.25, 2.25	a)	—	1 2.69		145)
A23 187	Ca²⁺	6	ligand A: —	a)	1 2.01	1 2.10	1 2.21		—	—	2:1 (L:M) "head-to-tail" dimer	142)
			ligand B: —	a)	1 1.92	1 2.02	1 2.22		—	—		
	Ca²⁺	7	ligand A: —	a)	1 2.27	1 2.37	1 2.69		1 (H₂O) 2.38	—	2:1 (L:M) "head-to-tail" dimer	143)
			ligand B: —	a)	1 2.28	1 2.38	1 2.58		—	—		

a) does not apply L = ligand, M = metal ion; Crystallographic coordinates are not available in cases where bonding distances are not given

Two different structures of an A23187 complex with calcium have been reported [142, 143]. In both of these, the Ca^{2+} ion is coordinated by two antibiotic anions each contributing a carboxy oxygen, a carbonyl oxygen, and the nitrogen of the benzoxazole ring system. The dimer is held together by coordination to the metal ion and by hydrogen bonds between the pyrrole nitrogen and the carboxy oxygen of the other A23187 ligand (see Fig. 20). The two dimeric forms, however, differ in the coordination geometry. In one complex form, the Ca^{2+} ion is sixfold coordinated [142] but in the second case it has sevenfold coordination [143]. The additional coordination site is occupied by a water molecule which is placed at the exterior of the complex. The fact that one complex form involves a water molecule at the surface while the other does not, suggests that the former is an intermediate in the complexation process, as pointed out by Smith and Duax [34].

Recently, a new divalent carboxylic ionophore called ionomycin has been isolated [112]. It was shown to be an effective calcium complexone [113] and to cause catecholamine release from rat pheochromocytoma cells [144]. Ionomycin is unique among the polyether antibiotics in chelating Ca^{2+} as a *dibasic* acid whereas all the other members of the nigericin family are *monobasic*. The dianion results from the ionization of the carbonyl and the enolized β-diketon moieties (see Fig. 15 for formula). A crystallographic study of several very similar forms of the Ca^{2+} and Cd^{2+} salts [145] revealed on octahedral coordination of the metal ion provided by one of the carboxylate oxygen atoms, both oxygens of the β-diketonate group, two hydroxy oxygens, and an ether oxygen of one of the tetrahydrofuran rings (Fig. 21). In the crystalline state, the complexes are joined in patris by two hydrogen bonds. The resulting dimers exhibit primarily lipophilic surfaces.

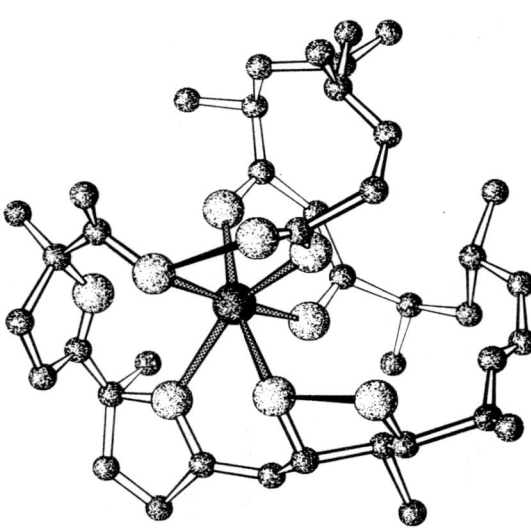

Fig. 21. Ca^{2+} complex of ionomycin. Hydrogen atoms have been omitted

Table 8. Uncomplexed polyether antibiotics and their adducts with chiral ammonium salts

Antibiotic	Number and lengths of internal "head-to-tail" hydrogen bonds	Comments	Ref.
monensin	1 2.66	The ligand shows several subtle changes of conformation as compared to the M^+ complexes. A water molecule occupies a site in the central cavity	123
grisorixin	1	The ligand displays the same conformation as in the Ag^+ complex. A water molecule takes the place of the metal ion	168
salinomycin	—	The p-iodophenacyl ester derivative was employed for X-ray analysis. The molecule adopts a helical structure lacking head-to-tail hydrogen bonding	169, 170
X-206	1	Conformation similar to that in the Ag^+ complex. A water molecule occupies the central cavity	156
septamycin	—	The p-bromophenacyl ester derivative was employed for X-ray analysis	171
A204A	1 2.99	The central cavity is occupied by a water molecule	172
lasalocid A	1	Refers to the 5-bromo derivative crystallized from non-polar solvents. The structure is a "head-to-tail" dimer enclosing a water molecule	135
lasalocid A	1 2.53	Refers to the free acid crystallized from methanol. Monomeric structure. A hydrogen-bonded methanol molecule is enclosed in the central cavity	138, 139
isolasalocid A	—	The compound does not exist in the characteristic cyclic conformation	173
A23187	1	The hydrogen bond is of the $N-H \cdots O$-type	140
lasalocid A 4-bromophenethylamine	1 2.75	1:1 Host-guest complex. The guest molecule is hydrogen-bonded to the antibiotic	130
X-14547A 4-bromophenethylamine	—	2:1 Host-guest complex. Only one of the antibiotic molecules is ionized	174, 175

5 Crown Ethers: Synthetic Macrocyclic Multidentates

5.1 Basic Stereochemistry of the 1,4-Dioxa Group in Polyether Complexes

As we have seen, X-ray studies of the ionophorous antibiotics and their cation complexes were able to explain many of the steric factors that determine the selectivity patterns shown by these ligands. However, more systematic investigations on the relationship between host-cavity size and guest-ion radius could only be carried out using simpler synthetic ligands as models. In 1967, Pedersen reported the synthesis and complexing properties of a new class of compounds named crown ethers [256] which are able to mimic effectively their natural counterpieces.

The most common crown ethers are shown in Fig. 22. Since application of IUPAC rules to polyethers leads to somewhat cumbersome designations, we will follow the simple "crown" nomenclature proposed by Pedersen [256].

Polyethers are built up from 1,4-dioxa units, $O-CH_2-CH_2-O$. The minimum energy conformation of these units is staggered with torsion angles about C-C bonds being synclinal (60°) and about C-O bonds being antiperiplanar (≈ 180 °C) (for definitions see Fig. 23). These preferences, however, do not preclude deviations if required by ring formation or cation complexation.

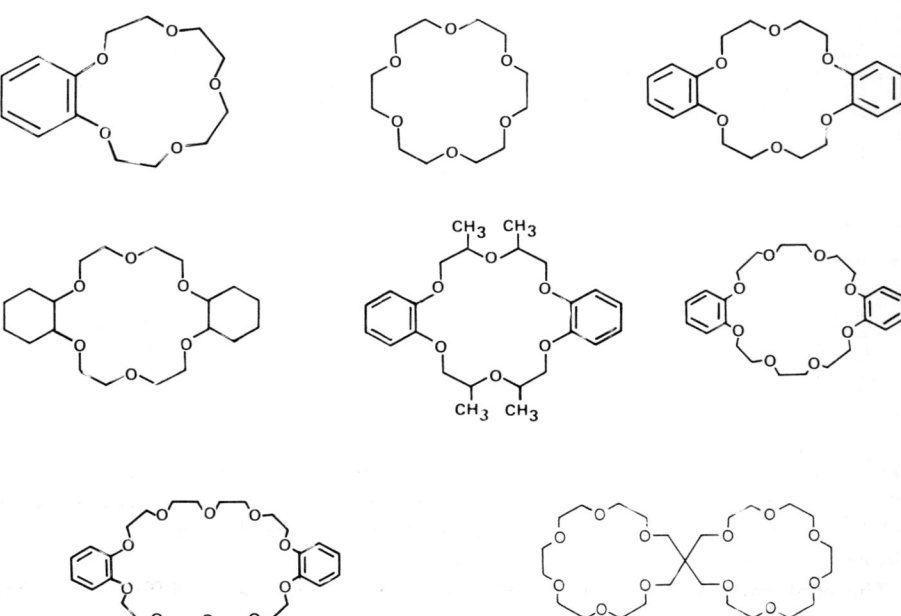

Fig. 22. Structural formulae of crown ethers. From left to right and top to bottom: benzo[15]crown-5, [18]crown-6; dibenzo[18]crown-6, dicyclohexano[18]crown-6; tetramethyldibenzo[18]crown-6, dibenzo[24]crown-8; dibenzo[30]crown-10, spiro-bis[19]crown-6

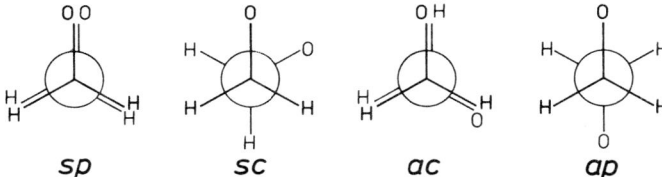

Fig. 23. Definition of torsion angles shown for the O—CH$_2$—CH$_2$—O unit. From left to right: *synperiplanar (sp), synclinal (sc), anticlinal (ac), antiperiplanar (ap)*

The latter process may cause major distortions of the torsion angles about C-O bonds whereas in the complexes, C-C bonds are essentially invariant in being synclinal ("all-gauche conformation") in order to place the ether oxygens close to the cation. It is noteworthy that if C-O bonds are forced into the synclinal conformation they usually adopt torsion angles larger than 70° to avoid short 1,4 CH ... HC contacts [257].

Interestingly, aliphatic C-C bonds are found to be systematically short in all the crystal structures of cyclic and open-chain oligoethers [240, 241, 330] and in their complexes with cations [244] as well as with neutral guests [340, 341]. Thus, the mean for these bond lengths is 1.507 Å in the uncomplexed hexaether 18-crown-6 [240] and even shorter by 0.03 Å in its Cs$^+$ complex [198] whereas the usually quoted value for a C(sp^3)-C(sp^3) bond is 1.537 Å [256]. Formerly, this effect was believed to be an artifact caused by inadequate treatment of thermal vibrations during the crystallographic refinement procedure [240]. Recently, however, Dunitz and coworkers repeated the structural analysis of 18-crown-6 at a temperature of 100 K so as to diminish thermal motions, and still, they found the mean C-C bond length (1.512 Å) to be only slightly increased with respect to the room temperature structure [241]. From this it may be concluded that the effect is at least partially real. A possible reason for the short C-C bonds might be the slightly polarized character of the adjacent C-O bonds which causes partially positive charges on the carbon atoms. By application of non-empiric SCF calculations to the hydrogen molecule it could be shown that the H-H bond length would slightly decrease if the protons were fractionally more positive than unity [342].

The KNCS complex of the hexaether [18]crown-6 [192] is highly ordered inasmuch as all the C—C—O—C and C—O—C—C torsion angles are close to 180° and all the O—C—C—O torsions are close to 60°. As a result, the average O ... O distance is as short as 2.82 Å which actually represents a Van der Waals contact. This unfavorable interaction is more than compensated for by the ion-dipole attraction induced by the complexed cation. The potassium ion is located at the center of the macrocycle and coordinated to the six ether oxygens which are all in a planar arrangement or, to be more exact, lie alternately above and below their mean plane by 0.2 Å (Fig. 24). Additionally, there is weak interaction between the metal ion and two disordered SCN$^-$ ions in the crystal lattice.

The conformation found in the K$^+$ complex of [18]crown-6 differs from that adopted by the free ligand, because in the uncomplexed molecule unfavorable alignment of dipoles is not compensated by a cation. Rather, the free coronand actually attains a somewhat elliptical conformation which allows for 1,5- and 1,8-CH ... O interactions [240, 241] (see Fig. 25). This is achieved by adjustment of several torsion angles including two C-O and two C-C bonds forced into synclinal (80°) and antiperiplanar (174°) conformation, respectively. As a result, not all the oxygen atoms point to the interior of the molecule as they do in the potassium complex. This, together with the elliptical shape of the molecule, is a general feature common to most of the uncomplexed polyethers.

The conformation of uncomplexed [18]crown-6 found in the crystalline state has been shown by IR spectroscopy to be predominant also in solution [258], and force-field calculations have essentially confirmed the same structure [241, 259]. Interestingly, it was found that the "ordered" D_{3d} [18]crown-6 conformation observed in the potassium complex would be lower in energy by 7 kJ · mol^{-1} were it not for the aforementioned unfavorable electrostatic interactions [259]. The tendency to remove the latter by complexation of a cation and thus enable the ligand to adopt a low-energy conformation might contribute considerably to the high affinity of [18]crown-6 to potassium ions.

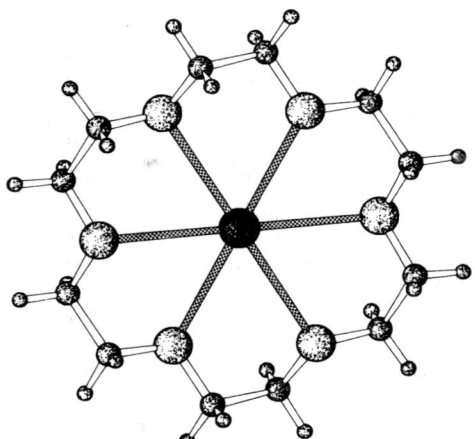

Fig. 24. Structure of the [18]crown-6 complex with potassium isothiocyanate. The anion is not shown

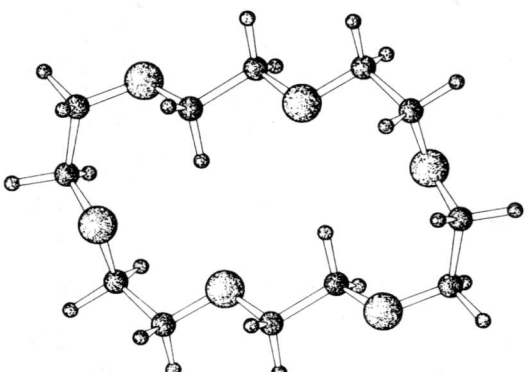

Fig. 25. Conformation of the free hexaether [18]crown-6. Note the elliptical shape of the molecule

5.2 The Ion-Cavity Concept

Examination of the diameters of ligand cavities and of the diameters of alkali and alkaline earth ions as given in Table 9 clearly shows that either the metal ion is too small to fill the cavity, or too large to fit in it, or it just meets the cavity size. The

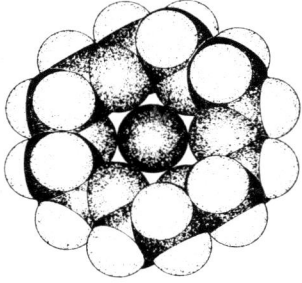

Fig. 26. Space-filling model of [18]crown-6-K$^+$

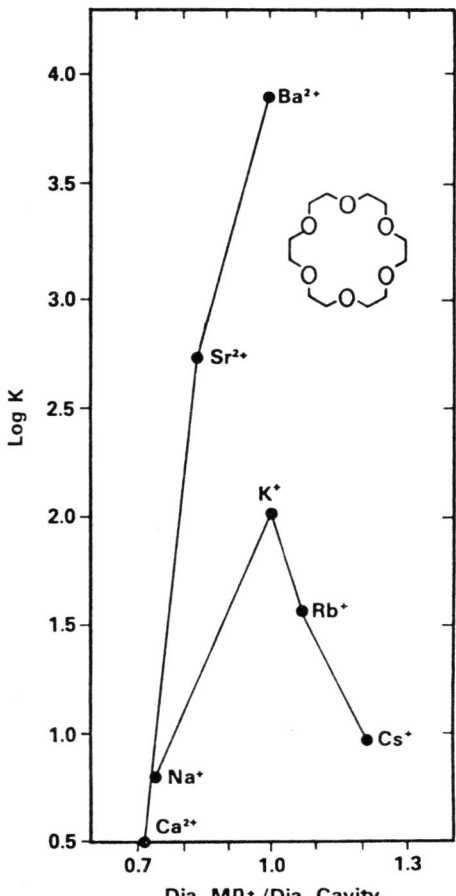

Fig. 27. Variation of the stability constant in water for [18]crown-6 complexes with alkali and alkaline ether cation depends on the degree of fit between host and guest. The ratio of cation size to crown cavity size is depicted on the abscissa (from Ref. [344])

Table 9. Diameters of crown cavities calculated from crystallographic data (after ref. [244]). Diameters of some alkali cations are given for comparison (all values in Å units)

[12]crown-4	1.2	Li^+	1.20
[15]crown-5	1.72–1.84	Na^+	1.90
[18]crown-6	2.67–2.86	K^+	2.66

latter is the case for [18]crown-6 and K^+, as can be seen from the space-filling model in Fig. 26, and in fact, potassium is the one most strongly complexed among the alkali metal ions [260]. The relationship between complex stability and the degree of fit between ligand and cation is depicted in Fig. 27.

Both Rb^+ and Cs^+ are too large to be accommodated into the 18-crown-6 cavity. Therefore, they occupy a site somewhat distant from the plane of the ether oxygens. Rb^+ is situated by 1.19 Å and Cs^+ even by 1.44 Å above this plane, which leads to a less favorable interaction with the ligand donor atoms, thus explaining the diminished stabilities of the [18]crown-6 complexes with these ions. The crown ether conformation does not alter in comparison to the K^+ complex. Coordination of the metal ion is completed by contacts with the bridging SCN^- ions, thus resulting in the formation of a 2:2 dimeric structure (Fig. 28) [197, 198].

On the other hand, Na^+ is too small to completely fill the ligand cavity. To render a sufficient interaction with all the possible ether oxygens, the ligand wraps around the sodium ion, one of the donor atoms thus occupying an apical position of the coordination sphere while the remaining five oxygens lie approximately in a plane. The metal ion is also coordinated to a water molecule. The SCN^- anion

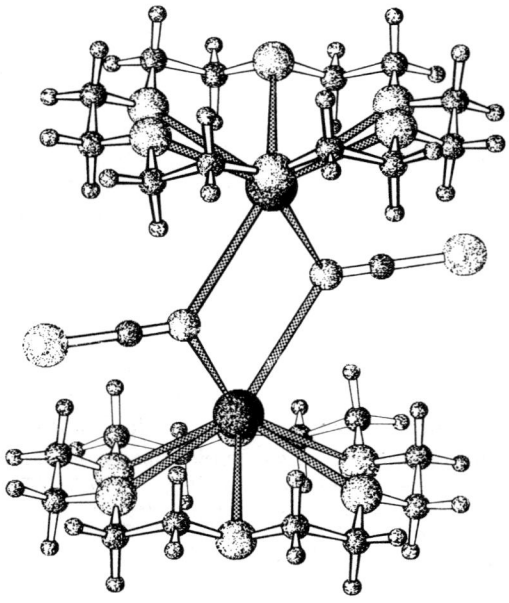

Fig. 28. Spatial structure of the 2:2 complex between [18]crown-6 and CsNCS

does not interact directly with the Na$^+$ but is hydrogen-bonded to the H$_2$O molecule (see Fig. 29). The perturbation of the crown-ether conformation is indicated by four torsion angles about C-O bonds severely deviating from the expected range [190].

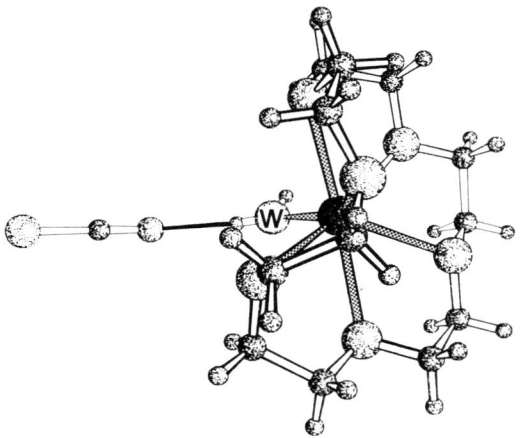

Fig. 29. The [18]crown-6 complex with NaNCS. "Double action" water molecule indicated by "W"

5.3 Effects of Anion and Cation Type on Complex Structure

Though the cavity-metal ion size relationship is very plausible, it should however be emphasized that this concept is oversimplified. Actually, there are other important factors ruling the complex structure such as ligand substituents and the anion involved. Thus, the ligand dibenzo[18]crown-6, which is less flexible than the unsubstituted hexaether, does not show any distortion in its complex with NaBr [208]. The sodium occupies a site slightly distant from the center of the macrocycle with contacts to the ether oxygens considerably longer than in the [18]crown-6-NaSCN complex (mean 2.71 Å, as compared to 2.55 Å). This is also true for a 1:1:1 complex between 18-crown-6, NaP(CN)$_2$, and tetrahydrofuran (THF). Its X-ray analysis [191] revealed two distinct complex units one of which consists of the macrocycle with Na$^+$ inside the cavity and the oxygens of two THF molecules occupying the apical coordination sites. The other also contains the sodium ion complexed inside the macrocyclic cavity, two dicyanophosphide ions coordinating to the metal ion *via* one cyano nitrogen. The latter unit is actually the first known example of a crown ether complex with negative overall charge. In both complex ions, [18-crown-6 · Na · (THF)$_2$]$^+$ and [18-crown-6 · Na · (P(CN)$_2$)$_2$]$^-$, the macrocycle attains the completely regular conformation also found in the [18]crown-6 complexes with K$^+$, Rb$^+$, and Cs$^+$. This can be explained by the availability of suitable solvent molecules or anions to occupy the apical coordination sites and thus avoiding unfavorable distortions of the ligand conformation. Evidence is added to this view by the fact that in the [18]crown-6 complex with Ca(NCS)$_2$ [199], the Ca^{2+} ion is located in the center of the (undistorted) crown though the ionic radius of Ca^{2+} ion is very similar to that of Na$^+$ (see

Table 1). This is due to the presence of *two* coordinating anions which lead to a sufficiently occupied coordination sphere of the metal ion.

If the anion is strongly coordinated to the metal ion to form an ion pair, it tends to pull the latter out of the crown ether ring. Thus, whereas K^+ is located exactly in the ether oxygen plane in the [18]crown-6 complex with KNCS, it is 0.9 Å above this plane, if the anion employed is the chelating ethyl acetoacetate [193]. As a result, the K ... O (crown) distances are considerably increased (2.83–3.02 Å) as compared to the KNCS complex (2.77–2.83 Å).

This effect is even more pronounced in the complex between benzo-15-crown-5 and calcium picrate [189]. In this structure, the metal ion though smoothly fitting into the cavity of the 15-membered ring, does not have any direct contact to the latter at all. Rather, it prefers chelation by the two picrate ions each of which provides its phenolate oxygen and an orthonitro oxygen. The remaining coordination sites are occupied by three water molecules. Two of these are hydrogen-bonded to adjacent crown ether molecules and thus stabilize the structure (see Fig. 30). Interestingly, though being uncomplexed, the polyether ligand adopts a conformation distinct from that of the free benzo[15]crown-5 [239], but similar to that in other benzo[15]crown-5 complexes with metal ions. From this it may be concluded that hydrogen bonding from water molecules is able to induce the same conformational

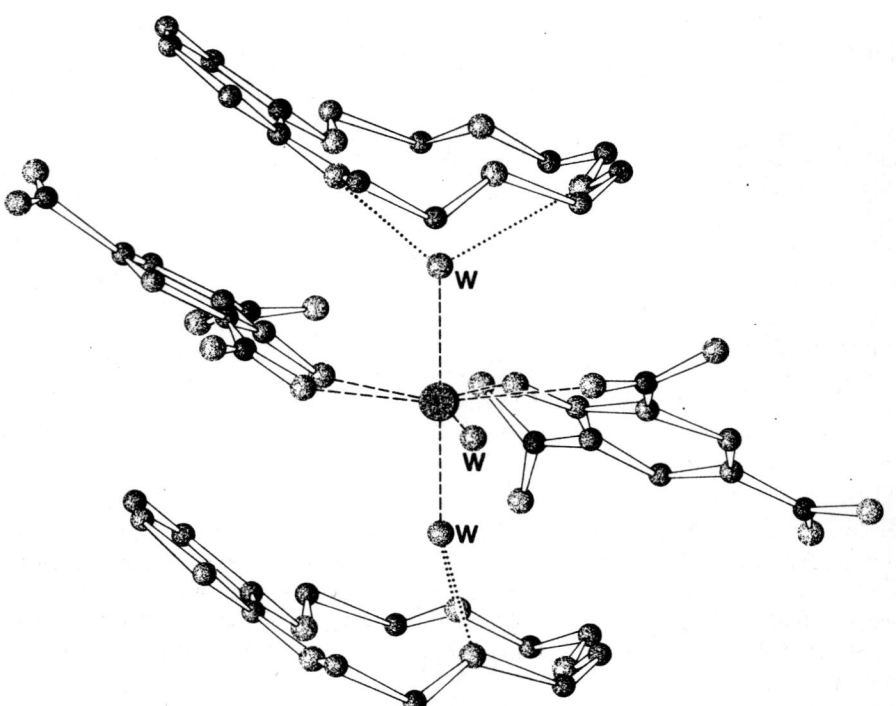

Fig. 30. Structure of the benzo[15]crown-5 complex with calcium picrate and three water molecules "W". Hydrogen atoms have been omitted for clarity

changes of the ligand as do metal ions, suggesting rather shallow energy barriers between different conformational states.

Clearlx, the above complex structure is dictated by the bulky picrate anions the o-nitro groups of which preclude further approach of the crown molecule toward the Ca^{2+} ion. When picrate is replaced by 3,5-dinitrobenzoate which lacks an o-nitro group, the Ca^{2+} is ninefold coordinated to the carboxylate oxygens and to all of the five crown ether oxygens but displaced by 1.38 Å from their mean plane [188]. Additionally, the crystal structure contains an uncomplexed molecule of benzo[15]crown-5 which is hydrogen-bonded to water.

The anion involved not only shows an effect on the distance between the metal ion and the ligand but also on the ligand conformation itself. Two crystal structures of a benzo[15]crown-5 complex with $Ca(NCS)_2$ have been published [186, 187], one being a water and the other the corresponding methanol solvate. The conformations of the cyclic ether in both these structures are very similar to each other, the Ca^{2+} being situated by 1.22 Å above the plane of the donor atoms. However, the ligand conformation differs considerably from that found in the complex with calcium 3,5-dinitrobenzoate [188]. This may be explained by the different coordination number (8 in the SCN^- complexes, and 9 in the 3,5-dinitrobenzoate complex) which influences the effective charge density of the cation, and in turn, together with the altered distance between the crown and the Ca^{2+}, results in a different polarization of the ligand by the cation. Furthermore, in the 3,5-dinitrobenzoate complex one of the bulky aromatic anions is stacked above the benzene ring of the ether.

The dependence of the complex structure on the anion involved is particularly pronounced in the case alkaline earth cations. Due to their relatively high charge densities (see Table 1), these ions are *always ion-paired* in their crown ether complexes. Only two exceptions to this rule are known: in the 12-crown-4 complexes with $MgCl_2$ [178] and $CaCl_2$ [179] the metal ions are heavily hydrated. These water molecules expel the chloride ions and, in the case of the Mg complex, also the crown molecule. The H_2O molecules display a "double action" [261] because they stabilize (1) the cation by coordination and (2) the anion by hydrogen bonding, thus diminishing the nucleophilic power of the latter.

Actually, highly nucleophilic anions such as Cl^- are deleterious for the synthesis of crown ether complexes, if they are not effectively stabilized by hydrogen bonding from protic solvents or other proton donors such as picric acid [262]. Anions that have been found useful for crown ether complex preparation are of soft HSAB character, e.g. SCN^-, ClO_4^-, Br^-, I^-, and picrate.

The difference in charge density between alkali and alkaline earth cations of similar radius, e.g. Na^+ and Ca^{2+}, causes some structural differences of their crown ether complexes which again demonstrate that the ion-cavity relation is not the only factor crucial for the complex structure. In the benzo-15-crown-5 complex with NaI [184], cation and anion are charge separated, only being connected by a H_2O molecule which again displays a "double action", as described above. The Na^+ ion lies only 0.75 Å above the crown, in comparison to 1.22 Å in the case of the corresponding complex with $Ca(NCS)_2$ [186, 187]. Moreover, there are severe differences in ligand conformation: One C—C—O—C torsion angle differs by as much as 90°.

5.4 Simultaneous Complexation of Metal Ion and Water or of two Metal Ions

Hitherto, we have seen two different structural possibilities for a crown ether complex with a cation the radius of which is too small to fill the ligand cavity: Either the ligand wraps around the ion as in [18]crown-6-NaSCN, or the crown is unchanged with the metal ion at its center assuming longer than optimum distances to the donor atoms as in [18]crown-6-NaP(CN)$_2$-THF. Yet another possibility is filling the host cavity by simultaneous complexation of a metal ion and a H$_2$O molecule. An example for this was recently found by Czugler and Kálmán in the structure of a lithium iodide complex of a multiloop crown ether in which two [19]crown-6 units are attached to each other by a spiro junction [232]. Each loop contains one Li$^+$ and one water molecule. The latter donates two hydrogen bonds to two ether oxygen atoms and interacts also with the lithium ion (Fig. 31). Besides to this water molecule, the metal ion is coordinated to three ether oxygens and to a second water molecule outside the ligand cavity. Thus, one of the ligand donor atoms interacts neither with H$_2$O nor Li$^+$, and it is just in this part of the coronand that large deviations from the normal torsion angles occur.

A somewhat similar situation has been found in the dibenzo-24-crown-8 complex with barium picrate [227]. The Ba^{2+} does only fill one compartment of the ring whilst the other is occupied by a water molecule which is hydrogen-bonded to the receptor and simultaneously coordinated to the Ba^{2+} ion. Interestingly, the situation is completely changed when ClO$_4^-$ is employed as anion instead of picrate. In this complex, the polyether wraps about the Ba^{2+} ion which is coordinated to all the ether oxygens and to the two perchlorate anions [226]. This is another example for the structurally decisive influence of the anion.

The ionic radius of K$^+$ is very close to that of Ba^{2+} but its charge density is much lower (see Table 1). As a result, dibenzo[24]crown-8 is able to form a complex with 1:2 stoichiometry (crown:metal ion) with KNCS in which *two* potassium ions

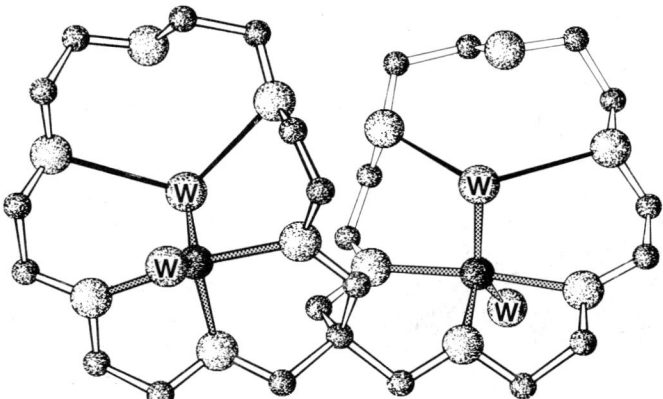

Fig. 31. 1:2 LiI complex of the bicyclic ligand spiro-bis[19]crown-6. Water molecules are indicated by "W". Neither iodide anions nor hydrogen atoms are included in the figure

are simultaneously complexed in the macrocyclic cavity and connected by bridging SCN⁻ ions [224, 225]. A somewhat similar arrangement was detected by X-ray crystallography in the 1:2 complex between the same ligand and sodium *o*-nitrophenolate [223] (Fig. 32).

Similar to dibenzo[24]crown-8, the much larger dibenzo[30]crown-10 complexes two Na⁺ ions simultaneously [228]. With potassium and rubidium, however, only 1:1 complexes are formed [229-231]. This is because the ligand completely wraps around

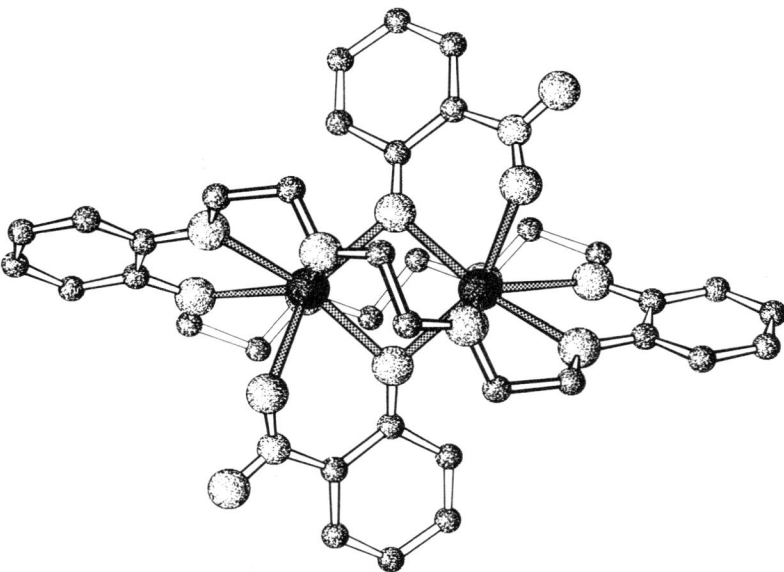

Fig. 32. Structure of the 1:2 dibenzo[24]crown-8 complex with sodium *o*-nitrophenolate. Hydrogens are not shown

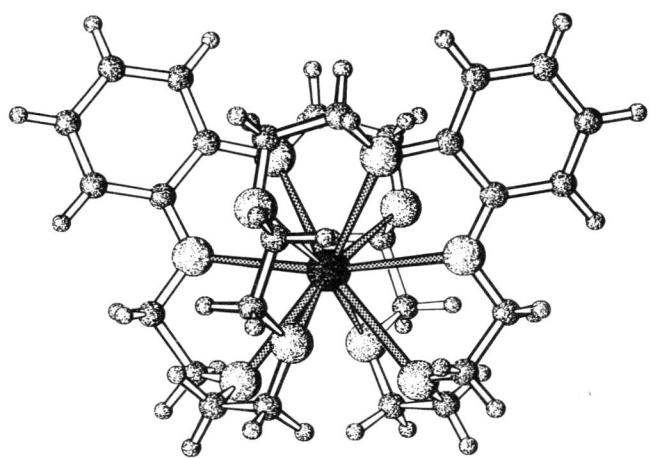

Fig. 33. Three-dimensional structure of the dibenzo[30]-crown-10 complex with K⁺

the complexed ion leading to a very effective three-dimensional encapsulation (Fig. 33) similar to that achieved by valinomycin.

5.5 Sandwich Formation

The cations Rb^+ and Cs^+ are too large in size for the [18]crown-6 cavity and lie above it. However, ^1H-, ^{13}C- and ^{133}Cs-NMR studies of the [18]crown-6- and dibenzo[18]crown-6-Cs^+ systems [263–266] indicated that at low temperatures formation of *2:1 sandwich complexes* occurs, but it should be emphasized that the anions involved in these investigations were BPh_4^- and BF_4^- which do not coordinate to cations and thus favor sandwich formation. Nonetheless, sandwich complexes are a common structural feature in many crown complexes with cations that are too large to fit into the macrocyclic cavity. The first crown sandwich whose structure was determined by X-ray crystallography was the KI complex of benzo[15]crown-5 [185]. In this complex, the potassium ion is simply coordinated by all of the ten ether oxygens, thus being completely shielded from the anion (Fig. 34).

The complex stoichiometry may, in some cases, be very sensitive to subtle changes in ligand configuration. Thus, the meso isomer "F" of 7,9,18,20-tetramethyldibenzo-[18]crown-6 (for formula see Fig. 22) forms a 2:2 dimer with CsNCS by bridging isothiocyanate ions [214], as was found in [18]crown-6-CsNCS itself. However, if the optically active isomer "G" is the ligand, use of the racemate leads to a 2:1 charge-separated sandwich in which Cs^+ is 12fold coordinated [214, 215].

Finally, it should be mentioned that sandwich formation is not a privilege of crown ether complexes with metal ions displaying ionic radii which are too large to fit into the ligand cavity. Infrared spectroscopic studies revealed that benzo[15]crown-5 forms a 2:1 sandwich with Na^+ (which could smoothly fit into the cavity), if the anion is tetraphenyl borate [267]. This is due to the inability of the anion to provide donor atoms for the Na^+ ion which thus requires a second crown for a sufficient coordination.

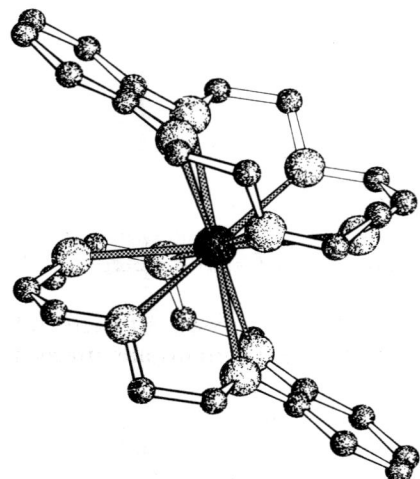

Fig. 34. 2:1 Sandwich-type complex between benzo-[15]crown-5 and K^+. For clarity, hydrogen atoms have been omitted

5.6 Ternary Crown Ether Complexes

In 1970, Pedersen reported on the formation of ternary adducts between dibenzo-[18]crown-6, potassium thiocyanate or iodide, *and* thiourea [268]. The stoichiometries varied between 1:1:1 and 1:1:6. One of these "supercomplexes", a 1:1:1 adduct between the crown ether, KI, and thiourea, has recently been the subject of an X-ray structural analysis [212]. It revealed that the potassium ion located in the center of the cyclic hexaether is coordinated to all the ether oxygens and to the iodide anion as was also found in the crystal structure of the simple 1:1 complex between the same ligand and KI [211]. Thiourea is not involved in the complexation of the metal ion, nor does it have any contact to the polyether ligand. Instead, it forms polymeric, hydrogen-bonded sheets. One hydrogen atom of each amide group is in contact with the sulfur of an adjacent thiourea molecule whereas the other one is involved in an N—H ... I hydrogen bond to the iodide anion (see Fig. 35). As a result of this involvement of the anion in hydrogen bonding, the K^+ ... I^- distance is slightly larger than the one found in the simple KI complex. Apparently, thiourea stabilizes the anion by hydrogen bonding as found for H_2O in many other crown ether complexes, thus facilitating complexation of the metal ion by the coronand. The resulting adduct might be called a *doubly wrapped salt* [268].

6 Macropolycyclic Host Molecules: Cryptands and their Cation Complexes

Macropolycyclic ligands, commonly referred to as *cryptands*, contain intramolecular cavities of three-dimensional shape (*"crypts"*). In their complexes (*"cryptates"*) with alkali and alkaline earth cations, they display considerably enhanced stabilities with respect to crown ethers (*"cryptate effect"*) [13, 289]. Thus, the K^+ complex of [2.2.2] cryptand (for nomenclature see Fig. 36) is by a factor of 10^5 more stable than the corresponding diaza[18]crown-6 complex and even by four orders of magnitude compared with the valinomycin potassium complex [293]. Furthermore, the smaller cryptands exhibit pronounced peak selectivity for alkali or alkaline earth cations (see Table 12) which agrees well with the ion-cavity size concept.

In bicyclic oligoethers which are usually designated as [2]-cryptates, two nitrogen atoms serve as bridgeheads. Each of these may be oriented either inward or outward with respect to the central cavity, leading to three possible stereoisomeric forms: *in-in*, *in-out*, and *out-out* (Fig. 37) [13]. Due to the electronic lone pairs on the nitrogen atoms pointing toward the metal ion, the most favorable isomer for complex formation is the *in-in* form which was actually found by X-ray crystallographic methods in all the cation complexes of cryptands and also for the uncomplexed [2.2.2] cryptand [291].

The *in-out* and *out-out* forms have been detected in the crystal structures of N-borane-[1.1.1] [292] and N,N-diborane-[2.2.2] [291] where BH_3 groups are attached to the amine nitrogens.

As in the case of monocyclic polyether ligands, the preferred conformations about C-C and C-O bonds are *synclinal* and *antiperiplanar*, respectively. The C—N—C—C, C—C—N—C torsion angles may lie in either of these low-energy ranges.

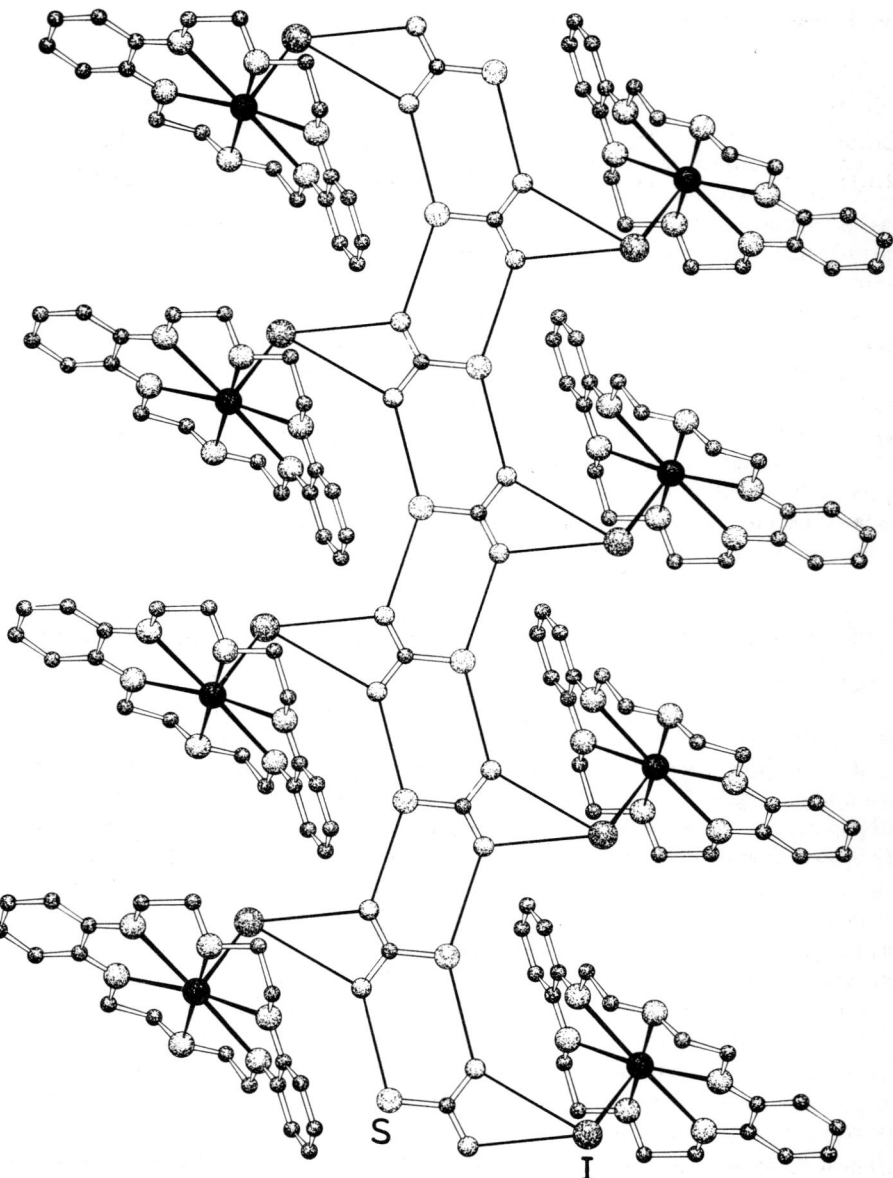

Fig. 35. Polymeric structure of the ternary "super" complex between dibenzo[18]crown-6, potassium iodide, and thiourea. Note the endless sheet of thiourea molecules. Iodide and sulfur indicated by I and S, resp.

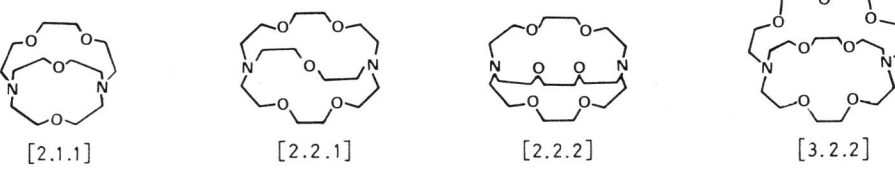

[2.1.1]　　　[2.2.1]　　　[2.2.2]　　　[3.2.2]

Fig. 36. Structural formulae of some cryptands

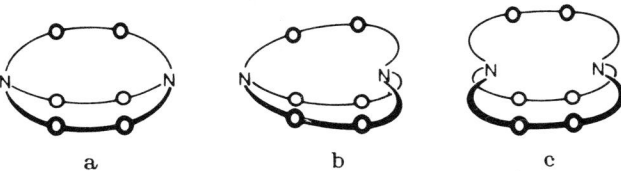

Fig. 37a—c. Cryptands: **a** *in-in*, **b** *in-out*, and **c** *out-out* stereoisomers (from Ref. [15])

A large number of [2.2.2] complexes have been subjected to X-ray analyses (see Table 13). The cation-ligand size relation for this cryptand has its optimum for potassium as the complexed species (cavity diameter ≈ 2.80 Å, ionic diameter 2.66 Å). However, the cryptand is flexible enough to accommodate the undersized Na^+ ion. This is accompanied by significant twisting of the ligand as expressed by some less favorable torsion angles [272]. Moreover, whereas Na^+ is fairly regularly eightfold coordinated in the [2.2.2]-NaI cryptate [272], somewhat different coordination modes have been detected in the crystal structure of ([2.2.2]Na)$_2$ Fe$_2$(CO)$_6$(μ_2-PPh$_2$)$_2$ [274] and ([2.2.2]Na)$_4$Sn$_9$ [277]. In these cases, the coordination number of sodium is decreased to seven and six, respectively. On the other hand, the twisting of the cryptand in both these structures is less pronounced. Apparently, an interplay of less favorable coordination of the guest ion and conformational strain is operative in [2.2.2] complexes with Na^+.

The flexibility of the [2.2.2] cryptand is also evident from its complexes with larger ions such as Rb^+ and Cs^+. If going from K^+ to Rb^+ to Cs^+, the distance between the two bridgehead nitrogen atoms which is significant for the cavity size increases from 5.75 to 6.00 to 6.07 Å, accompanied by a change of the mean torsion angle about C-C bonds from 54 to 67 to 70° [279]. Moreover, a N ... N separation as short as 4.92 Å has been found in [2.2.2]-Ag^+ [293], whereas it reaches its maximum of 6.87 Å in the free host molecule [291]. The latter is of an elongated shape as described in Chapter 5 for uncomplexed crown ethers.

In the [2.2.2] complex with cesium, the Cs^+ ... O distances are somewhat smaller than the corresponding sum of the Van-der-Waals radius of oxygen and the ionic radius of Cs^+, as was similarly observed in the Cs^+ complex of nonactin (see Chap. 3). This indicates that the cation presses heavily on the ligand. In fact, it was concluded from ^{133}Cs-NMR data that in solution, [2.2.2] forms an *exclusive complex* with Cs^+, i.e. the metal ion does not occupy the central ligand cavity but is only partly

51

Table 10. Complexes of cyclic crown ethers

Compound	Coord. no.	Bond distances M–O	M-anion	M–O (solv.)	Stoichiometry and type of structure	Ref.
[12]crown-4 complexes						
[Na · L$_2$]$^+$ [Cl · (H$_2$O)$_5$]$^-$	8	2.47–2.52	—	—	2:1 sandwich	176)
[Na · L$_2$]$^+$ [OH · (H$_2$O)$_8$]$^-$	8	2.44–2.51	—	—	2:1 sandwich	177)
L · [Mg(H$_2$O)$_6$]$^{2+}$ (Br)$_2^-$	6	—	—	2.05–2.08	Mg^{2+} not encapsulated in the crown	178)
[Ca · L · (H$_2$O)$_4$]$^{2+}$ (Cl$^-$)$_2$	8	2.51–2.54	—	2.38–2.40	1:1; Ca^{2+} 1.63 Å above the crown plane	179)
[Cu · L · Cl$_2$]	6	2.11–2.40	2.21, 2.23	—	1:1	180)
[(UO$_2$) · L · (H$_2$O)$_2$]$^{2+}$ (NO$_3^-$)$_2$	6	1.77–2.05	—	2.37, 2.37	1:1; UO$_2^{2+}$ encapsulated in the crown	181)
[15]crown-5 complexes						
[Ba · L$_2$]$^{2+}$ [Br$_2$(H$_2$O)$_2$]$^{2-}$	10	2.75–2.88	—	—	2:1 sandwich; crowns are apparently disordered	182)
L · [Cu(H$_2$O)$_2$Br$_2$]	4	—	2.35, 2.37	1.95, 2.00	Cu^{2+} not encapsulated in the crown	183)
benzo[15]crown-5 complexes						
[Na · L · H$_2$O]$^+$ I$^-$	6	2.35–2.43	—	2.29	1:1; I$^-$ hydrogen-bonded to water	184)
[K · L$_2$]$^+$ I$^-$	10	2.77–2.96	—	—	2:1 sandwich	185)
[Mg · L · (NCS)$_2$]	7	2.17–2.20	2.06	—	1:1	186)
[Ca · L · (NCS)$_2$ · H$_2$O]	8	2.46–2.61	2.42, 2.43	2.40	1:1; Ca^{2+} 1.23 Å above the crown	186, 187)
[Ca · L · (NCS)$_2$ · MeOH]	8	2.51–2.55	2.40, 2.49	2.38	1:1; very similar to the former complex	186, 187)
[Ca · L · (dinitrobenzoate)$_2$] · [L · (H$_2$O)$_3$]	9	2.52–2.78	2.46–2.47	—	one of the two crowns does not interact with Ca^{2+}	188)
L · [Ca(pic)$_2$(H$_2$O)$_3$]	7	—	2.28–2.59	2.24–2.38	Ca^{2+} not encapsulated in the crown	189)

[18]crown-6 complexes

Complex	CN				Notes	Ref
[Na·L·H$_2$O]$^+$ NCS$^-$	7	2.45–2.62	—	2.32	1:1; water displays a double action	190)
[Na·L·(THF)$_2$]$^+$	8	2.71–2.79	—	—	first example of a negatively charged crown unit	191)
[Na·L·(P(CN)$_2$)$_2$]$^-$	8	2.73–2.78	2.44, 2.48	2.36, 2.36		
[K·L]$^+$ NCS$^-$	6(8)	2.77–2.83	3.19, 3.19	—	1:1; very weak interaction with the disordered anion	192)
[K·L·(ethyl acetoacetatoenolate)]	8	2.83–3.02	2.65, 2.73	—	1:1; K$^+$ 0.9 Å above the oxygen plane	193)
[K·L·(tosylate)]	8	2.78–2.94	2.69, 2.93	—	1:1; crown disordered	194)
[K·L·(H$_2$O)$_2$]$^+$	8	2.78–3.05	—	2.78, 2.82	1:1; both the K$^+$ ions displaced from the mean crown planes by 0.92 and 0.78 Å, resp.	195)
[K·L·(H$_2$O)·MoO$_4$]$^-$	8	2.76–2.99	2.79	2.81		
[K$_2$·L$_2$·(H$_2$O)$_2$·Mo$_6$O$_{19}$]	8	2.72–2.88	2.70, 2.72	2.89, 2.93	2:2; two crown-K$^+$ units linked by the anion	196)
[Rb·L·(NCS)]$_2$	8	2.93–3.15	3.23, 3.31	—	2:2; formation of dimers by bridging SCN$^-$ anions	197)
[Cs·L·(NCS)]$_2$	8	3.03–3.27	3.30, 3.32	—	2:2; bridging SCN$^-$ ions	198)
[Ca·L·(NCS)$_2$]	8	2.56–2.74	2.35, 2.35	—	1:1; Ca^{2+} in the center of the ether oxygen plane	199)
(L)$_2$·[Mn(H$_2$O)$_6$]$^{2+}$ (ClO$_4^-$)$_2$	6	—	—	2.14–2.22	Mn^{2+} not encapsulated in the crown ether	200)
(L)$_2$·[Co(H$_2$O)$_6$]$^{2+}$ [CoCl$_4$]$^{2-}$	6/4	—	—	2.04–2.12	both the Co^{2+} ions not complexed by the crowns	201)
([UCl$_3$·L]$^+$)$_2$·[UO$_2$Cl$_3$(OH)(H$_2$O)]$^{2-}$	9	2.48–2.61	—	—	1:1; the published coordinates are suspicious	202)
L·(H$_2$O)$_3$·[UO$_2$(H$_2$O)$_2$(NO$_3$)$_2$]	8	—	2.48–2.49	2.43, 2.43	UO$_2^{2+}$ not encapsulated by the crown ether	

benzo(L')- and 4-nitrobenzo(L'')/[18]crown-6 complexes

Complex	CN				Notes	Ref
[Rb·L'·(NCS)]$_2$	8	2.91–3.13	3.04, 3.05	—	2:2; formation of dimers by bridging SCN$^-$ ions	204)
[Sr·L'·(H$_2$O)$_3$]$^{2+}$ (ClO$_4^-$)$_2$	9	2.66–2.72	—	2.55–2.58	1:1	205)
[Ba·L'·(H$_2$O)$_2$·(ClO$_4$)$_2$]	10	2.80–2.85	2.79, 2.94	2.78, 2.84	1:1	205)
[Rb·L''·(NCS)]	8	2.95–3.08	2.90	—	1:1; Rb$^+$ interacts with the nitro group of an adjacent ligand molecule	206)

Table 10. (continued)

Compound	Coord. no.	Bond distances M—O	M-anion	M—O (solv.)	Stoichiometry and type of structure	Ref.
[Cs · L'' · (NCS)]	8	3.04–3.25	Cs–N: 3.44 Cs–S: 3.68	—	1:1; Cs^+ interacts with nitro group as in the former complex	207)
dibenzo[18]crown-6 complexes						
$NaBr \cdot L \cdot 2 H_2O$: Molecule A: $[Na \cdot L \cdot H_2O \cdot Br]$ Molecule B: $[Na \cdot L \cdot (H_2O)_2]^+ Br^-$	8 8	2.54–2.89 2.63–2.82	2.82 —	2.35 2.27, 2.31	1:1; two distinct complexes with different coordination. A is a complexed ion pair and B a complexed cation	208)
$[Na \cdot L]^+ NCS^-$	6	2.74–2.89	—	—	1:1; occupancy of cation site is 45% for Na^+ and 55% for Rb^+ (see below)	209, 210)
$KI \cdot L \cdot 1/2 H_2O$: molecule A: $[K \cdot L \cdot I]$ molecule B: $[K \cdot L \cdot H_2O]^+ I^-$	7 7	2.73–2.79 2.73–2.79	3.52 —	— 2.72	1:1; two distinct complex molecules with different modes of coordination	211)
$[K \cdot L \cdot I \cdot (thiourea)]$	7	2.71–2.80	3.57	—	1:1:1 complex between KI, crown, and thiourea; polymeric	212)
$[Rb \cdot L \cdot (NCS)]$	7	2.86–2.94	2.94	—	1:1; occupancy of cation site is 55% for Rb^+ and 45% for Na^+ (see above)	209, 210)
$[Sm \cdot L \cdot (ClO_4)_3]$	10	2.41–2.59	2.36–2.64	—	1:1	213)
tetramethyldibenzo[18]crown-6 complexes (isomer "F" = L': methyl groups cis, anti, cis cis; isomer "G" = L'': methyl groups trans, anti, trans, trans)						
$[Cs \cdot L' \cdot (NCS)]_2$	8	3.07–3.34	3.30, 3.32	—	2:2; formation of dimer by bridging SCN^- ions	214)
$[Cs \cdot (L'')_2]^+ NCS^-$	12	3.12–3.36	—	—	2:1 charge-separated sandwich	214, 215)

dicyclohexanol[18]crown-6 complexes					
[Na · L · (H$_2$O)$_2$]$^+$ Br$^-$	8	2.68–2.97	—	2.35, 2.35	1:1; ligand is the cis-anti-cis isomer	216, 217)
[Ba · L · (NCS)$_2$ · (H$_2$O)]	9	2.80–2.91	2.88, 2.88	2.80	1:1; ligand is the cis-syn-cis isomer	218)
[La · L · (NO$_3$)$_3$]	12	a)	a)	—	1:1; ligand is the cis-syn-cis isomer	219)
([UCl$_3$ · L]$^+$)$_2$ UCl$_6^{2-}$	9	2.47–2.65	—	—	1:1; cis-syn-cis isomer	220)
2,3-naphtho[20]crown-6 complex					
[K · L · (NCS)]	7	2.73–2.88	3.26	—	1:1; K$^+$ interacts only weakly with disordered SCN$^-$	221)
4,18-dioxo bezo[21]crown-7 complex					
[K · L · (NCS)]	8	2.77–3.07	N: 2.80 S: 3.49	—	1:1; K$^+$ interacts also with S of adjacent SCN$^-$	222)
dibenzo[24]crown-8 complexes					
[Na$_2$ · L · (o-dinitrophenolate)$_2$]	6	2.47–2.62	2.30–2.40	—	1:2; each Na$^+$ coordinated to 3 ether oxygens	223)
[K$_2$ · L · (NCS)$_2$]	7	2.73–2.98	2.87, 2.88	—	1:2; both the SCN$^-$ bridge metal ions	224, 225)
[Ba · L · (ClO$_4$)$_2$]	10	2.76–3.04	2.72, 2.79	—	1:1	226)
[Ba · L · (H$_2$O)$_2$ · (pic)$_2$]	10	2.86–3.00	2.67–3.09	2.73, 2.77	1:1; ligand cavity occupied by Ba^{2+} *and* a water molecule	227)
dibenzo[30]crown-10 complexes					
[Na$_2$ · L · (NCS)$_2$]	7	2.40–2.59	2.36	—	1:2	228)
[K · L]$^+$ I$^-$	10	2.85–2.93	—	—	1:1	229)
[K · L]$^+$ NCS$^-$	10	2.84–2.96	—	—	1:1	230)
[Rb · L]$^+$ [(NCS) · (H$_2$O)]$^-$	10	2.96–3.19	—	—	1:1	231)
spiro-bis[18, 18′-19-crown-6] complex					
[Li$_2$ · L · (H$_2$O)$_4$]$^{2+}$ (I$^-$)$_2$	5	1.93–2.21	—	1.93, 1.96	1:2; each loop contains a Li$^+$ and a water molecule	232)

Table 13. Coordination distances found in cryptates by X-ray crystallographic studies. The N ⋯ N separation is also given to demonstrate the relative flexibility of the cryptand cages

Compound	Coord. no.	Bond distances (Å)				N⋯N separation (Å)	Comments, if any	Ref.
		M–O	M–N	M-anion	M–OH_2			
[2.1.1]-complexes								
$[Li \cdot L]^+ I^-$	6	2.08–2.17	2.29, 2.29	—	—	4.21		269)
[2.2.1]-complexes								
$[Na \cdot L]^+ SCN^-$	7	2.45–2.52	2.59, 2.70	—	—	4.94	Na^+ occupies a central position	270)
$[K \cdot L \cdot (NCS)]$	8	2.70–2.87	2.90, 2.92	2.78	—	5.14	K^+ lies in 18-membered ring	270)
$[Co \cdot L]^{2+} [Co(SCN)_4]^{2-}$	7	2.10–2.22	2.20, 2.24	—	—	4.20	Co^{2+} occupies a central position	271)
[2.2.2]-complexes								
$[Na \cdot L]^+ I^-$	8	2.57–2.58	2.72, 2.78	—	—	5.50		272)
$([Na \cdot L]^+)_2 [Fe(CO)_4]^{2-}$	8	av. 2.53	av. 2.80	—	—	av. 5.60		273)
$([Na \cdot L]^+)_2$ $[Fe_2(CO)_6 (\mu_2 - PPh_2)_2]^{2-}$	7	2.50–2.75	2.69, (3.14)[a]	—	—	5.84		274)
$[Na \cdot L]^+ Na^-$	8	av. 2.57	av. 2.72	—	—	5.43		275)
$([Na \cdot L]^+)_3 Sb_7^{3-}$	8	2.40–2.71	2.83, 2.94	—	—	av. 5.84		276)
$([Na \cdot L]^+)_4 Sm_9^{4-}$:								
Molecule A:	6	2.47–2.57	2.90, 2.93	—	—	5.92		277)
Molecules B, C, D:	8	2.41–2.76		—	—	6.07		
$[K \cdot L]^+ I^-$	8	2.78–2.79	2.87, 2.87	—	—	5.75		278)
$[Rb \cdot L]^+ [(SCN)(H_2O)]^-$	8	2.88–2.93	2.99, 3.01	—	—	6.00	water hydrogen-bonded to SCN^-	279)
$[Cs \cdot L]^+ [(SCN)(H_2O)]^-$	8	2.96–2.97	3.02, 3.05	—	—	6.07	water hydrogen-bonded to SCN^-	279)
$[Ca \cdot L \cdot (H_2O)]^{2+} (Br^-)_2$	9	2.49–2.55	2.72, 2.72	—	2.42	5.44	water displays a double action	280)

Complex						
[Ba·L·(NCS)(H$_2$O)]$^+$ NCS$^-$						
Molecule A:	10	2.75–2.82	2.94, 3.00	2.91	2.88	5.94	In both molecules, H$_2$O displays a double action	281)
Molecule B:	10	2.74–2.89	2.99, 3.00	2.88	2.84	5.99		
[Tl·L]$^+$ [(HCOO)(H$_2$O)]$^-$	8	2.90–2.91	2.95, 2.95	—	—	5.89	water hydrogen-bonded to HCOO$^-$	282)
[Pb·L·(NCS)(SCN)]	10	2.73–2.98	2.86, 2.91	N: 2.64 S: 3.12	—	5.76		283)
([La·L·(NO$_3$)$_2$]$^+$)$_3$ [La(NO$_3$)$_6$]$^{3-}$	12	2.64–2.74	2.81–2.85	2.63–2.69	—	av. 5.62		284)
[Eu·L·(ClO$_4$)]$^{2+}$(ClO$_4^-$)$_2$	10	2.44–2.52	2.64, 2.70	2.67, 2.71	—	5.34	one ClO$_4^-$ acts as a bidentate anion	285, 286)
[Sm·L·(NO$_3$)]$^{2+}$	10	2.44–2.57	2.75, 2.78	2.48, 2.50	—	5.53	NO$_3^-$ acts as a bidentate	287)
[Sm(NO$_3$)$_5$(H$_2$O)]$^{2-}$								
[Ag·L]$^+$ [Ag$_3$(SCN)$_4$]$^-$	8	2.66–2.85	2.40–2.50	—	—	4.92	Ag–N bonds are partially covalent	293)
[3.2.2]-complexes								
[Ba·L·(H$_2$O)$_2$]$^{2+}$(NCS$^-$)$_2$	11	2.80–3.09	3.08, 3.18	—	2.81, 2.87	6.10	water molecules display a double action	288)

L = ligand; refers to the cryptand quoted in the corresponding headline
a) not considered a binding distance

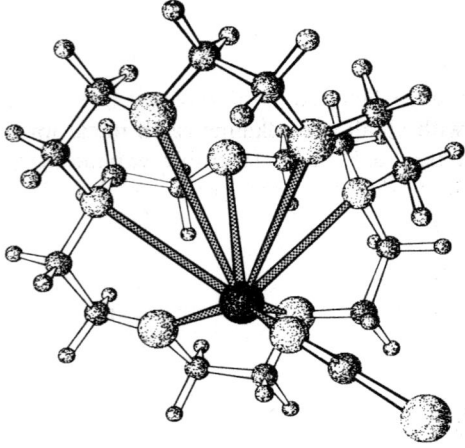

Fig. 38. Structure of the "exclusive" [2.2.1] complex with potassium isothiocyanate

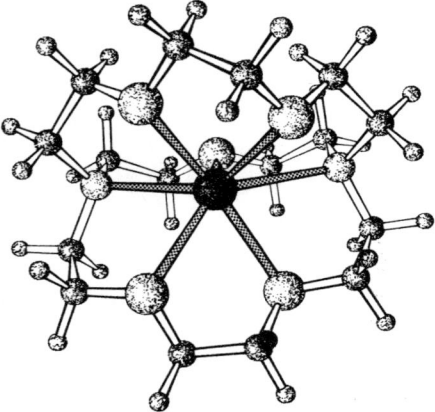

Fig. 39. The "inclusive" [2.2.1] —Na$^+$ complex

thermodynamic studies revealed that this selectivity is of entropic rather than enthalpic origin [296]. This suggests that the ion-cavity radius concept though being more generally valid for cryptands than for crown ethers, again is only one criterion among others that determine the complex stability.

7 Open-Chain Polyethers: Wrapping of Metal Ions

7.1 Podands with Aromatic Donor End Groups

When going from the cyclic [18]crown-6 to its open-chain analog *pentaglyme* (Fig. 40) as a ligand for K$^+$, the complex stability decreases by factor as high as 10^4 [324] although both these ligands offer the same number of donor atoms. The enhanced stability of cyclic crown ether complexes with respect to those of corresponding linear polyethers (*podands*) is attributed to a *macrocyclic effect* [325-328] which is most likely of entropic origin [326, 327].

Because of this, open-chain polyethers were believed until recently not to be capable of forming crystalline complexes with alkali and alkaline earth metal ions. However, in 1977 Vögtle and coworkers reported on considerable enhancement of complex stabilities by attaching rigid aromatic donor end groups to the oligo-(ethylene glycol) backbone as in podands *1–10* (Fig. 41) [14, 329]. A series of rubidium iodide complexes with ligands *1–4* of increasing length containing 8-quinolinole moieties as rigid end groups has been subjected to X-ray analyses [330]. As a rule, it was found that *all* the heteroatoms of the ligands are coordinated to the Rb$^+$ ion. Furthermore, the steric preferences described in Chapters 5 and 6 for coronand and cryptand complexes are valid also for podands, i.e. torsion angles about C—O bonds are generally *antiperiplanar* (ap), those about C—C bonds *synclinal* (sc). Again, deviations from this rule reflect special steric requirements.

The short ligand *1* (five heteroatoms) wraps about the Rb$^+$ ion in a circular arrangement with all the torsion angles in the usual range [297]. While the cation is shielded

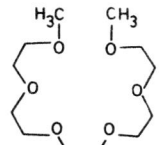

Fig. 40. Pentaglyme

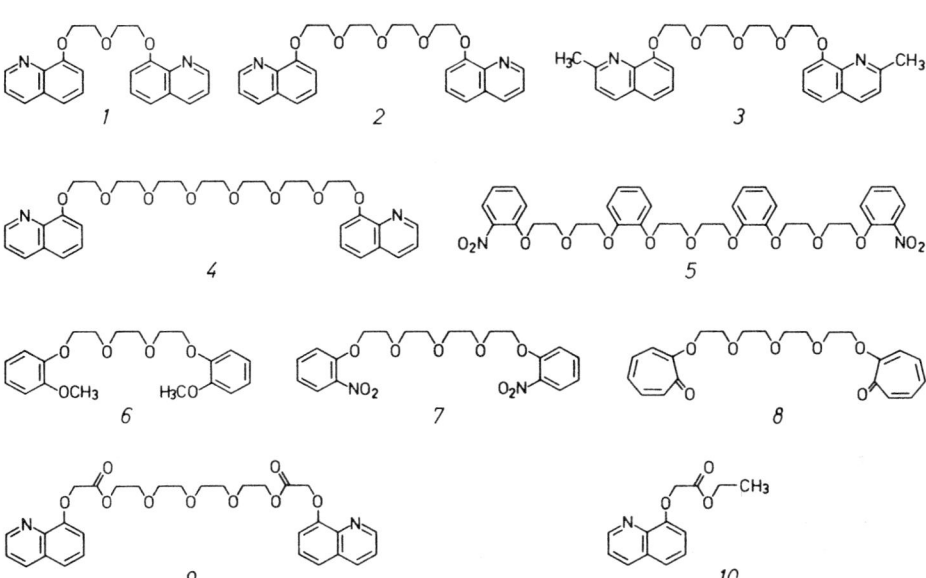

Fig. 41. Structural formulae of synthetic open-chain polyether ligands containing aromatic donor end groups

by the oligoether from one side, it is free to coordinate to two (crystallographically equivalent) iodide anions on the opposite side (Fig. 42). If podand *1* is extended by two ethylene glycol units as in *2*, it does not fit circularly around the rubidium ion but has to adopt a helical structure [298, 299]. Thus, this *achiral* ligand forms a *chiral* complex. Starting from one heteroaromatic moiety, the polyether wraps around the Rb$^+$ ion in a planar arrangement, the metal ion being positioned 0.75 Å above this plane. However, in order to avoid intramolecular collision between the two quinoline systems, a sharp turn (indicated by an arrow in Fig. 43) is introduced in the smooth wrapping of the ligand around the equatorial coordination sphere of the cation. This "kink" is achieved by a rotation of one C—O torsion angle from

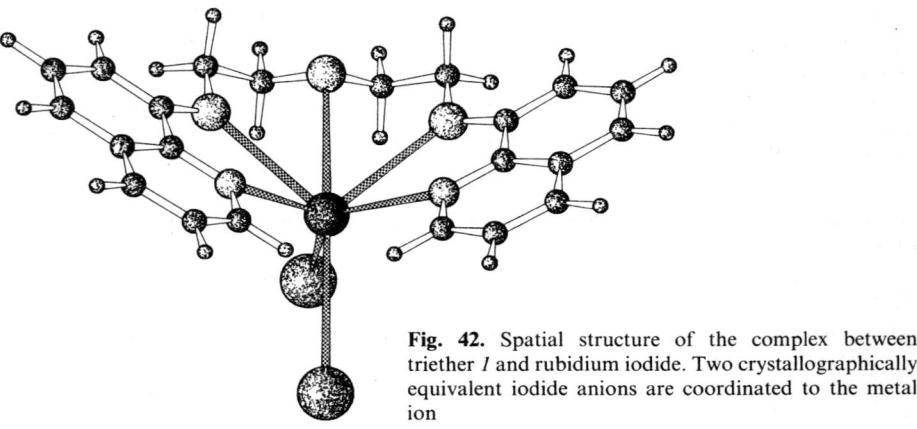

Fig. 42. Spatial structure of the complex between triether *1* and rubidium iodide. Two crystallographically equivalent iodide anions are coordinated to the metal ion

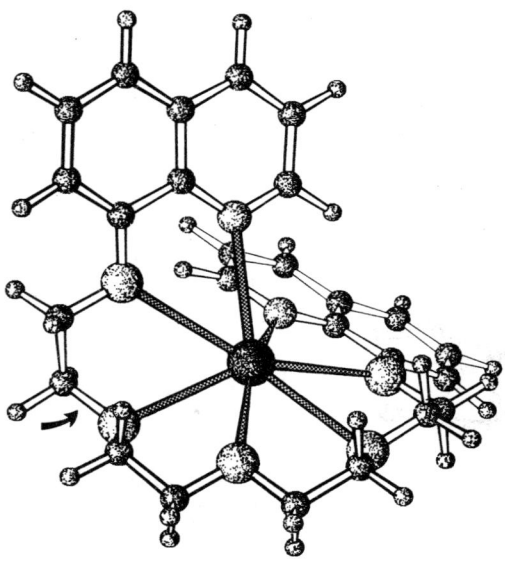

Fig. 43. Helical structure of the Rb$^+$ complex of podand *2*. The ligand "kink" with *sc* instead of *ap* orientation of torsion angle C—O—C—C is indicated by an arrow

antiperiplanar to synclinal. As a result, the nitrogen atom of the second heteroaromatic moiety is placed by 3 Å above the equatorial ligand plane.

Interestingly, the $Rb^+ \ldots O$ distances to the *aromatic* oxygen atoms are significantly longer (3.07 and 3.09 Å) than those to sp^3 oxygens. This is due to the diminished basicity of the former [298], probably associated with an unfavorable orientation of the electron lone pairs with respect to the metal ion.

If quinaldine rather than quinoline moieties are introduced as ligand end groups, an entirely different complex structure results. Podand *3* displays a helical configuration with both heterocycles stacked parallel to each other at 3.4 Å [301] (Fig. 44). The helix is somewhat more continuous than in the quinoline complex and does not display an abrupt kink. Apparently, this is accomplished by slight deviations of *all* the torsion angles from their ideal values, which is particularly true for the aliphatic C—C bonds, the mean torsion angle about which is 71° (range: 65–75°) as compared to 62° (range: 60–67°) in the quinoline complex. In order to facilitate π—π interaction between the heterocycles, the torsion angle about one of the two "aromatic" C—O bonds is in the anticlinal (128°) rather than in the more common antiperiplanar range.

The interaction between Rb^+ and the aromatic nitrogens (distances 3.04 and 3.11 Å) is weaker compared to the quinoline complex (2.93 and 2.96 Å). This is compensated for by additional coordination to the iodide anion and by the energy gained from stacking interactions.

A plausible explanation for the structural differences between the RbI complexes with ligands as similar to each other as the quinoline pentaether *2* and its methylated analog *3* is evident if one assumes the rigid donor end groups to be potential centers of nucleation of the binding process [331]. In the case of 8-quinolinole this is particularly striking as this compound itself is a strong chelating ligand for alkali and alkaline earth cations [332]. This view is supported by the short distances displayed

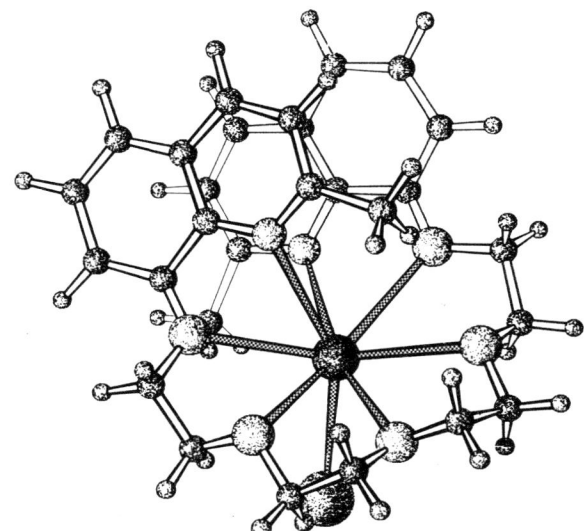

Fig. 44. Spatial structure of the complex between quinaldine ligand *3* and RbI

between Rb^+ and the quinoline nitrogen atoms which are actually even shorter than the sum of the corresponding atomic and ionic radii.

In contrast, 8-hydroxyquinaldine cannot function as "recognizable" site for the metal ion. The latter is not able to approach the donor atom sufficiently because of the presence of the 2-methyl group as reflected by the significantly elongated Rb^+ ... N distances.

range [302, 303]. All the heteroatoms may be considered coordinating if Rb^+ ... N

Ligand *4* with ten heteroatoms wraps smoothly around the Rb^+ ion with only one C—O torsion angle (127°) deviating largely from the expected antiperiplanar and Rb^+ ... O distances of 3.37 and 3.15 Å, respectively, are tolerated. The cation is completely shielded by the ligand (see Fig. 45) and therefore is unable to contact the anion.

Employing spectrophotometric titrations, Tümmler et al. [331] found noncyclic polyether ligands containing aromatic donor end groups to be strong but relatively nonselective host molecules for alkali cations; thus, in the case of ligand *2* log K was found to be 3.22, 3.51, and 3.06 for complexation of Na^+, K^+ and Rb^+, respectively. The authors explained this behavior by the high flexibility of the polyether chain which is able to easily adjust to the size of the metal ion. However, a recent X-ray study of the *2* · KSCN complex suggests another explanation [300]. Since K^+ is significantly smaller in size compared to Rb^+, one might expect the kink in ligand conformation, i.e. the deviation from planar surrounding of the cation, to occur one or two bonds earlier than in the Rb^+ complex. This, however, is not the

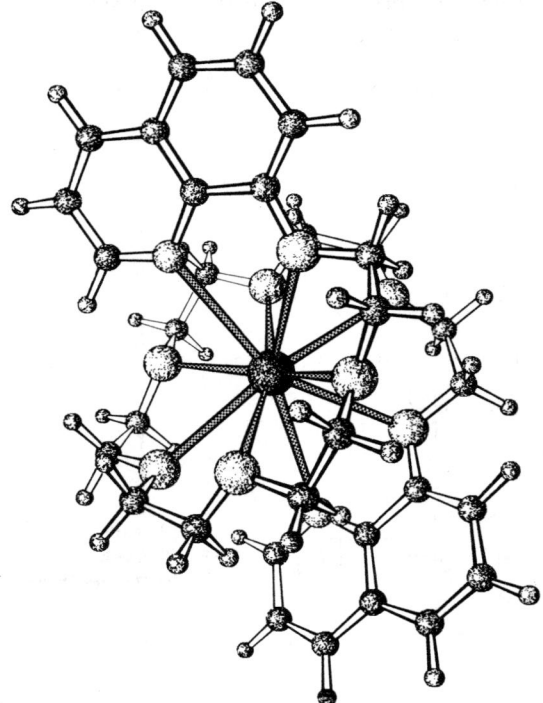

Fig. 45. Spherical wrapping of a metal ion: Podand *4*-Rb^+

case. Rather, the abnormal torsion angle is actually connected with the *same* C—O bond as in the Rb⁺ complex, and its value is almost identical with that found in the latter (75° as compared to 69°). Furthermore, though smotthly fitting into the cavity made up by the planar part of the ligand, the potassium ion is still placed 0.66 Å above it, i.e. it occupies a site nearly identical to that of the Rb⁺ ion. As a result, the K⁺ ... O distances are in the range 2.80–2.93 Å (average 2.86 Å), which is significantly *longer* than those observed in dibenzo[18]crown-6-K⁺ (2.71–2.80 Å, mean 2.76 Å) where the potassium ion lies within the plane formed by the ether oxygens [211,212]. On the other hand, the K⁺ ... N distances (mean 2.81 Å) are slightly *shorter* than the theoretical value (2.83 Å). This is accounted for by the strong interaction with both the terminal groups, and, in fact, these donor sites mainly determine the complex stabilities rather than the overall conformation of the polyether chain. Thus, the low selectivity of ligands of this type is obviously due to the predominant donor ability of the heteroaromatic end groups whereas the discriminating function of the ligand pseudo-cavity as complexing site is lost.

7.2 Complexes of Podands Containing End Groups Capable of Hydrogen Bonding

The ligands discussed in Section 7.1. are sometimes quoted as model compounds for polyether antibiotics (see Chap. 4). However, this does not exactly meet the situation because they lack the intramolecular hydrogen bond of the head-to-tail type which is characteristic of the polyether bioionophores. Furthermore, the latter

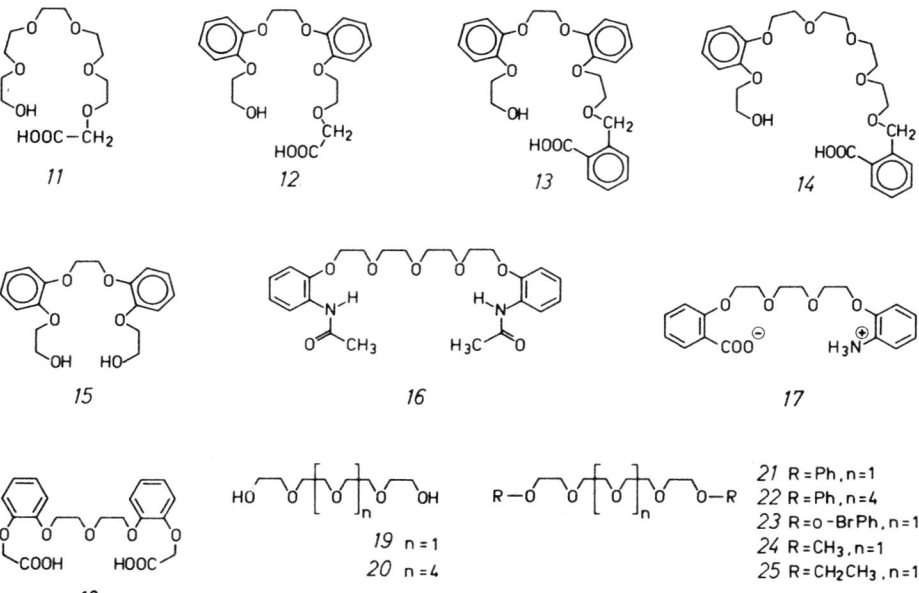

Fig. 46. Structural formulae of linear polyethers lacking rigid donor end groups

do not display helical structures in their complexes, as is observed with many synthetic podands.

Yamazaki et al.[333] were able to show that the hydrogen bonding is obviously a prerequisite to the ability of these ligands to transport metal ions across (artificial) membranes. These authors investigated the carrier properties of the open-chain polyethers depicted in Fig. 46 and found that *12* selectively transported K$^+$, as did *13* and *14* for Rb$^+$, whereas *11* and *15* were unable to transport alkali metal ions. This is in agreement with X-ray structural studies of both the NaNCS and KNCS complexes of *15* [314] which revealed that they do not contain the internal hydrogen bond which is essential for membrane transport. Rather, *two* ligand molecules are involved in the coordination of one metal ion in both complexes. As a result, the NaNCS complex is polymeric but of 1:1 stoichiometry whereas the KNCS complex consists of distinct 2:1 (ligand: metal ion) units (Fig. 47).

Although not containing rigid donor end groups as the ligands discussed in Section 7.1., these podands are still capable of forming alkali metal ion complexes due to their hydroxy end groups. In all the structures highlighted here, the latter display a *"double action"* similar to that described for water molecules in Section 5.3., i.e. they are coordinated to the cation and simultaneously stabilize the anion by hydrogen bonding, for example of the O—H ... N—C—S$^-$ type. This interaction prevents the anion from interacting too strongly with the metal ion and therefore, these polyethers containing hydroxy or carboxy groups are called *"double action ligands"* [261].

Recently, the first X-ray study on a metal ion complex of a synthetic oligoether ligand containing an internal head-to-tail hydrogen bond was completed[311]. The α-carboxy-ω-amino tetraether *17* (see Fig. 41 for formula) surrounds the Na$^+$ ion in a more or less planar arrangement. The terminal groups, COO$^-$ and NH$_3^+$,

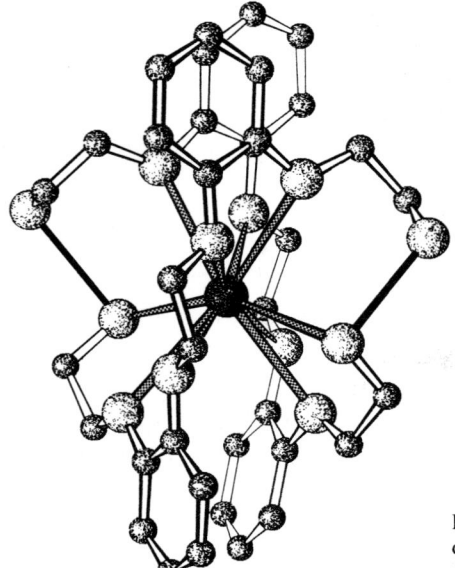

Fig. 47. Structure of the 2:1 complex between dihydroxypodand *15* and K$^+$. Hydrogen atoms are not shown

are linked to each other by a strong intramolecular N—H ... O hydrogen bond (Fig. 48). Presumably, the latter is to a much greater extent responsible for the pseudocyclic structure of the ligand than are the ion-dipole forces between Na$^+$ and the ether oxygens. This is also reflected by the fact that only three of the four ether oxygens contribute to the coordination of the sodium ion. A similar situation has been found, for example, in the Tl$^+$ complex of the polyether antibiotic lonomycin [153, 154]. ^{1}H-NMR studies suggest that the N—H ... O hydrogen bond is also present in the free ligand leading to a pseudocyclic configuration as well [334]. This is exactly what is found in naturally occurring polyether antibiotics so that it seems justified to consider *17* as a suitable analog of the latter, especially for the pyrrole ethers A23187 and X-14547 A both of which contain an intramolecular N ... H—O hydrogen bond. But still, there are important structural differences.

In the crystalline state, the synthetic α-carboxy-ω-amino ligand forms a 2:2 dimer by formation of hydrogen bonds with the ClO$_4^-$ anions (see Fig. 48). This arrangement is somewhat similar to that found in the 2:2 lasalocid complex with Na$^+$ [134] if one neglects the presence of the perchlorate anions.

Unfortunately, most of the uncomplexed podands are oils so that crystallographic studies are not possible. The only free linear polyether whose structure was elucidated

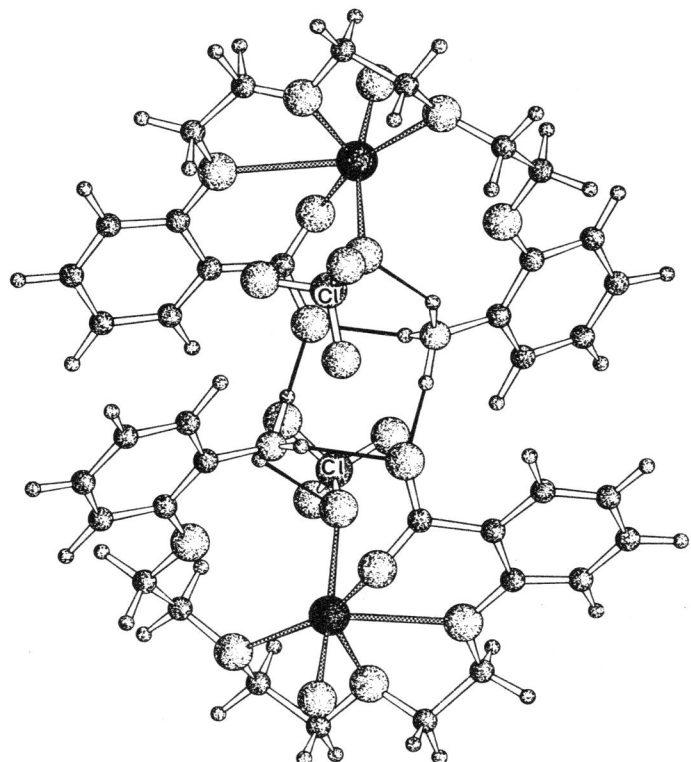

Fig. 48. 2:2 Dimer of the NaClO$_4$ complex of the α-carboxy-ω-amino ligand *17*

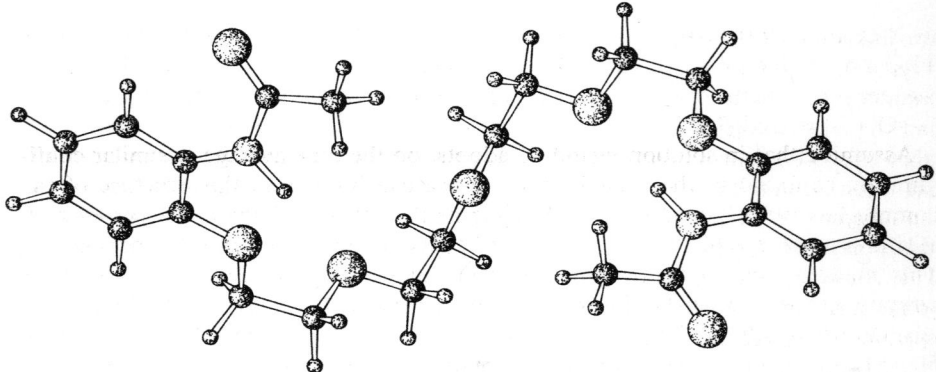

Fig. 49. S-like configuration of the uncomplexed amide ligand *16*

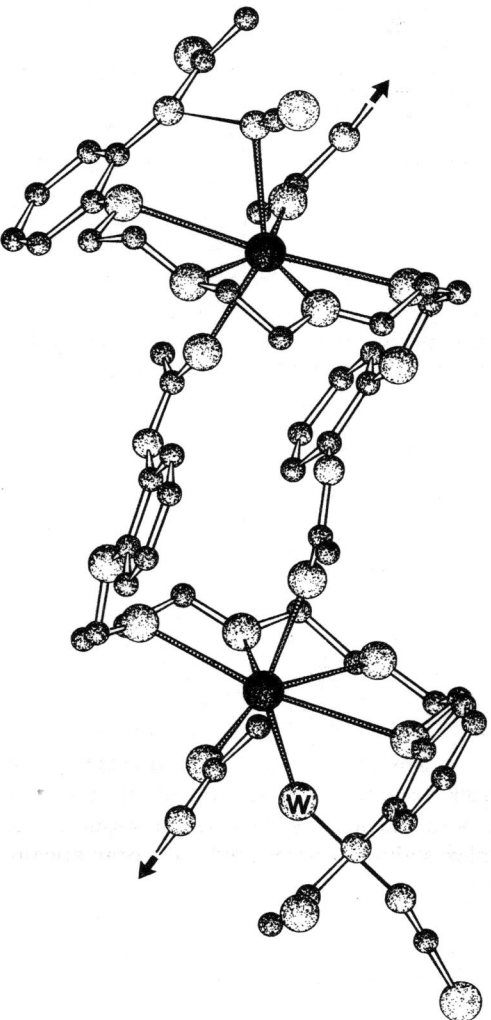

Fig. 50. Polymeric structure of the complex between polyether *16* and potassium isothiocyanate. Hydrogen atoms have been omitted for clarity. "W" indicates a water molecule. Continuation of adjacent molecules indicated by arrows

by X-ray methods is the amide ligand *16* [310] (for formula see Fig. 46). The molecule adopts an S-like configuration as is shown in Fig. 49. The torsion angles follow the general scheme being ($\pm$) synclinal for O—C—C—O but antiperiplanar for C—O—C—C and C—C—O—C.

Assuming that in solution ligand *16* adopts, on the time average, a similar configuration as found in the crystal structure, we can infer from the structure of the complex between *16* and K^+ that binding of the cation is initiated by recognition of one of the loops of this "empty" ligand by the cation, followed by coordination. This might be associated with a major conformational adjustment to allow for wrapping of the remaining donor atoms around the K^+ in a helical manner. These structural changes of the ligand are obvious from an inspection of torsion angles along the oligoether chain: Whereas C—O torsions remain antiperiplanar and those about C—C bonds synclinal, some of the latter change sign [310]. From this it can be concluded that rotation about C—C bonds contributes to the free activation energy $\Delta G^{\ddagger}$ of complex formation whereas unfavorable conformations around C—O bonds induced by sterical strain, if present, affect the free reaction enthalpy ΔG°.

The *16* · KNCS complex forms a complicated polymeric structure as depicted in Fig. 50 [310]. An important feature of this complex is the classical "double action" displayed by a water molecule which is hydrogen-bonded to an SCN^- anion, thus shielding it from the cation. The other isothiocyanate anion accepts a hydrogen bond from the amide N—H grouping.

The formation of this polymer is in good agreement with the high entropy of complexation ($\Delta S^{\circ} \approx -200$ J · K^{-1} · mol^{-1}) found for the corresponding $NaClO_4$ complex by ^{23}Na-NMR techniques [335]. In fact, polymerization is a process accompanied by large negative entropy changes. This agreement between the results provided by crystallographic and spectroscopic studies suggests that the polymeric structure revealed in the crystalline state is probably also present in (concentrated) solutions.

7.3 Linear Polyethers without Donor End Groups

The previous two sections suggest that either strong aromatic donors or protic groups capable of displaying double action have to be employed as end groups for a potent noncyclic oligoether ligand. Recently, however, Vögtle and his coworkers succeeded in obtaining crystalline *alkaline earth* complexes of ligands without any terminal donor groups, including the very simple *glymes* (glycol dimethyl ethers) [336].

Ligands of this type are of great interest, as they, though forming much less stable complexes than podands containing donor end groups, exhibit considerably enhanced *selectivities* with respect to the latter [331]. Thus, the ethylene glycol diphenyl ether *21* shows the highest peak selectivity for K^+ of all the podands hitherto tested. Substitution of one or both phenoxy groups by the 8-quinolinol moiety results in highly increased complex stabilities associated with dramatically diminished selectivity, as can be seen from Fig. 51.

The low stabilities of the oligo (ethylene glycol) diphenyl ether complexes are accounted for by unfavorable entropy changes upon complexation which are no longer important after attachment of aromatic donor end groups to the ligand

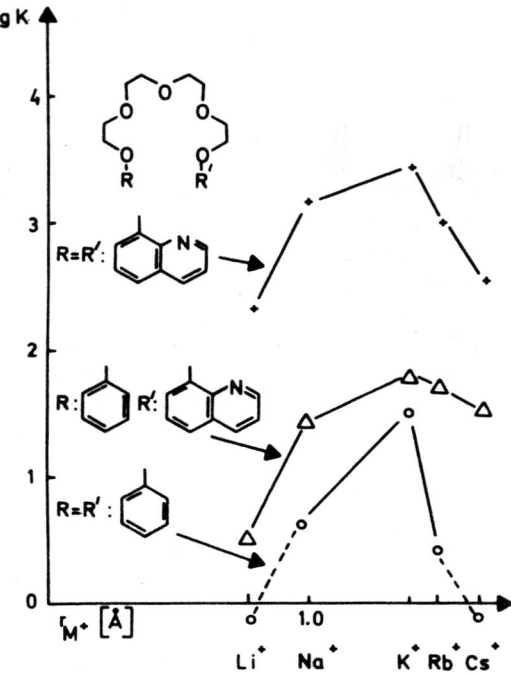

Fig. 51. Stabilities of complexes between open-chain pentaethers and alkali metal ions: Enhancement of stabilities by attaching quinoline moieties as donor end groups (from Ref. [331])

chain [331]. Until recently, this behavior together with the pronounced selectivity was not explainable. However, the first X-ray crystallographic study of a complex between a polyether lacking any donor end groups and a metal ion [317] provided much insight in the structural features that govern the selectivity pattern. Interestingly, the *octa*-ether 22 behaves only as a *hepta*dentate in its Ba(NCS)$_2$ complex (Fig. 52).

As found for the polyethers containing aromatic donor groups, the ligand wraps around the ion in a helical manner including a region of a more or less planar arrangement of the donor atoms in the equatorial zone of the coordination sphere. However, whereas the former ligands display *one* "kink" introduced by *one* abnormal torsion angle if necessary to avoid collision of the terminal chain segments, the structure described here contains *two* of these turns involving *four* C—O torsion angles being shifted to the *anticlinal* range. These conformational abnormalities, which are indicated by arrows in Fig. 52, locate the two phenyl groups below and above the mean ligand plane, respectively and suggest the following mechanism of cation inclusion [337].

Substitution of the Ba^{2+} solvation shell starts with the central ligand oxygens, and the steric hindrance of both the terminal segments of the ligand chain is overcome by a *cooperative mechanism*. Both the chain ends escape each other by undergoing unfavourable changes in conformation. This may be described by the term "binding of central portion followed by binding of floppy ends". In contrast, the mechanism postulated for podands containing donor end groups [298,331] is different: Starting from one end group, the ligand wraps around the metal ion

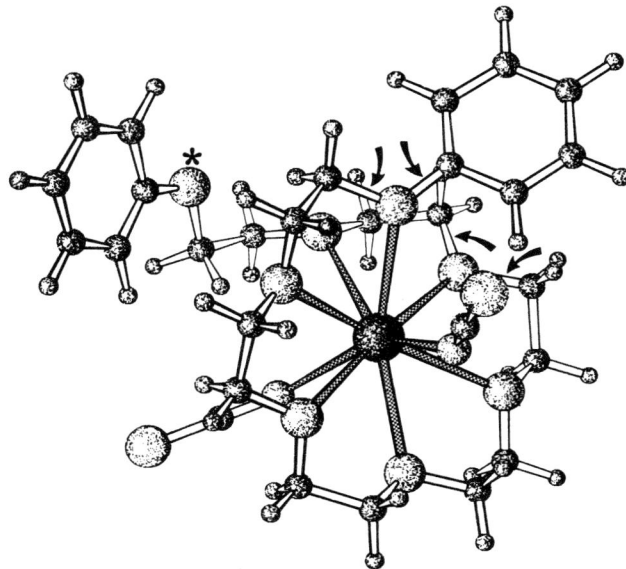

Fig. 52. X-ray structure of a Ba(NCS)$_2$ complex of the octaether *22*. Arrows indicate conformational changes from normally preferred *ap* into *ac* in the polyether chain. One of the oxygen atoms (marked by an asterisk) is not coordinated to the Ba^{2+} ion

in a regular arrangement until continuation of this process in hampered by intramolecular collision which forces the ligand into an unusual conformation. This is best designated as "binding of one end followed by binding of the other end".

Clearly, the fixation of the very flexible ligand *22* in a sterically disadvantageous conformation is entropically unfavorable. Furthermore, since the metal ion is located at a position coplanar with the central donor atoms, this ligand conformation is very sensitive to a change of the cation radius, thus providing an explanation for the pronounced peak selectivities exhibited by ligands without donor end groups.

7.4 Tripod Ligands

In 1977, it was found that the attachment of additional ligand atoms to linear polyethers resulted in a large enhancement of complex stabilities [338]. These *polypodands* exhibit a cryptand-like behavior also in their phase-transfer properties.

The first tripodand structure elucidated by X-ray crystallography was the KNCS complex of ligand *26* (see Fig. 53 for formula) [322]. The cation was found to be located in the center of a three-dimensional cavity and coordinated to all the available donor atoms (Fig. 54), reminiscent of prey cought by a three-legged octopus. From this structure, the designation of this tripodand as a "noncyclic cryptand" [338] appears to be justified.

This does not hold strictly, however, for tripodand *27* complexed with RbI. The metal ion is too big to be located in the central cavity. Therefore, it occupies a site in a two-dimensional cavity built up by two of the three octopus arms and distances to the donor atoms of the third arm are significantly longer [323]. Additionally, there is an interaction with the anion. As a whole, the binding situation is reminis-

Table 14. Geometrical features of podand complexes in the solid state

Compound	Coord. no.	Bond distances M—O	M—N	M-anion	M—OH$_2$	Comments, if any	Ref.
Complexes of podands containing rigid donor end groups							
[*1* · RbI]	7	3.06–3.17	2.97, 2.97	3.69, 3.90	—	Rb$^+$ coordinated to two crystallographically equivalent I$^-$ ions	297)
[*2* · Rb]$^+$ I$^-$	7	2.89–3.08	2.93, 2.96	—	—	helical structure	298, 299)
[*2* · K]$^+$ SCN$^-$	7	2.80–2.93	2.81, 2.82	—	—	helical structure	300)
[*3* · RbI]	8	2.84–3.12	3.05, 3.11	3.63	—	helical structure	301)
[*4* · Rb]$^+$ I$^-$	10	2.95–3.12	3.16, 3.36	—	—	helical structure	302, 303)
[*5* · K$_2$ · (SCN)$_2$]	8	2.69–3.18	a	—	—	K$^+$ interacts weakly with SCN$^-$	304, 305)
[*6* · Na(NCS)]	7	2.35–2.54	a	2.33	—	helical structure	306)
[*7* · K(NCS)]	8	2.79–3.22	a	2.72	—	helical structure	307)
[*8* · RbI]	8	2.94–3.18	a	3.67	—	planar structure	308)
[*9* · Ba(NCS)$_2$]	11	2.79–3.17	2.92, 2.92	2.87, 2.87	—	helical; carbonyl oxygens do not coordinate	309)
[*10* · Ba(NCS)$_2$]	9	2.76–3.01	2.89, 2.92	2.87, 2.91	2.73	2:1 complex; carbonyl oxygens are coordinated to the cation	309)
Complexes of podands containing hydroxy, carboxy, or amide end groups							
[*15* · Na(NCS)]	7	2.42–2.57	a	2.41	—	polymeric structure	314)
[(*15*)$_2$ · K]$^+$ SCN$^-$	10	2.86–3.05	a	—	—	2:1 complex	314)
16 · KSCN:							
Molecule A:							
[*16* · K · (H$_2$O)]$^+$ NCS$^-$	7	2.71–3.09	—	—	2.79	polymeric structure	310)
Molecule B:							
[*16* · K · (NCS)]	7	2.61–2.97	—	2.84	—		
[*17* · Na · ClO$_4$]$_2$	6	2.25–2.74	—	2.38, 2.41	—	polymeric structure	311)
([*18* · K]$^+$)$_2$ (pic$^-$)$_2$	8	2.73–2.90	a	—	—	dimeric structure	312, 313)
[*19* · Ca · (pic) (H$_2$O)]$^+$ pic$^-$	8	2.39–2.50	a	2.33, 2.62	2.38	water and ligand itself display a double action	315)
[*20* · Sr(NCS)]$^+$ SCN$^-$	9	2.56–2.73	a	2.57	—		316)

Structural Chemistry of Natural and Synthetic Ionophores and their Complexes with Cations

Complexes with podands containing no donor end groups					
[22 · Ba(NCS)$_2$(H$_2$O)]	10	2.79–3.11	a	2.77, 2.81	3.08	only 7 of the 8 ligand oxygens coordinated to Ba^{2+}	317)
[23 · (HgBr$_2$)]	7	2.72–3.06	a	(2.39, 2.41)	—		318)
[24 · (HgCl$_2$)]	7	2.78–2.96	a	(2.29, 2.31)	—		319)
[24 · (CdCl$_2$)$_2$]	6	2.41–2.74	a	2.42–2.68	—	two Cd^{2+} ions connected with each other by bridging Cl anions	321)
[25 · (HgCl$_2$)$_2$]	6	2.66–2.91	a	2.38	—		320)
Complexes of tripod ligands					
[26 · K]$^+$ SCN$^-$	10	2.77–3.00	2.94	—	—		322)
[27 · RbI]	8	3.03–3.06	3.00–3.20	3.76	—		323)

a does not apply; pic = picrate; The ligand numbering refers to the formulae depicted in Figs. 41 and 46

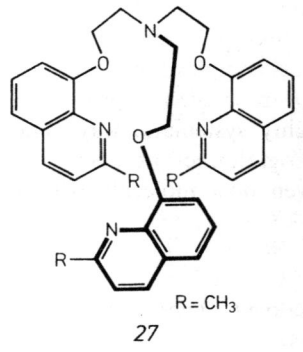

26

27 R = CH₃

Fig. 53. Structural formulae of tripod ligands

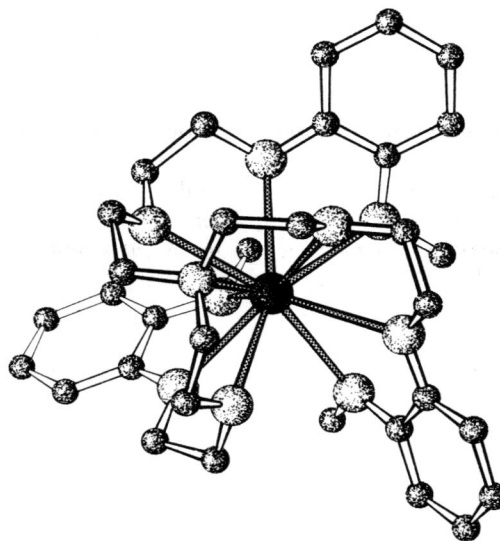

Fig. 54. Spatial structure of the 26-K⁺ complex

cent of the "exclusive" cryptates, e.g. [2.2.1] · KNCS, which we have discussed in Chap. 6.

For the sake of completeness, mention should be made of a water complex of the same tripod ligand 27 studied by X-ray diffraction recently [339]. The water molecule is complexed by only one of the three ligand loops employing two N ... H—O hydrogen bonds.

8 Concluding Remarks

Looking backwards, one has to agree that it is indeed mainly the results obtained from X-ray crystallographic studies that the present knowledge on the structural foundations of selective complexation is based on. What is needed now is refinement of this knowledge toward the goal of being able to explain the mechanisms of host-guest complex formation with metal ions in greater detail than is possible today.

A promising approach to this is the application of X-ray crystallography in a highly systematic way. This is to say, emphasis should not be put on the investigation of as many new ligands as possible but rather on complexes of a given host molecule with various guest entities and, possibly, crystallized from solvents of different polarity. From a precise knowledge of the conformational changes of the ionophore molecule it is possible to propose detailed mechanisms for both ion capture and release, as we have seen in Chapter 4 for monensin and lasalocid. These studies should, of course, be complemented by spectroscopic, thermodynamic and kinetic investigations.

9 Acknowledgement

Most of the figures were drawn by the UNIVAC 1100/82 computer of the Gesellschaft für wissenschaftliche Datenverarbeitung (GWD), Göttingen, employing the plot program SCHAKAL which was written by Dr. Egbert Keller, Freiburg, and generously placed at our disposal. We are grateful to the GWD operators for their efforts in obtaining high-quality plots.

We would like to thank Drs. Galena Tishchenko, Moscow, and Matyas Czugler, Budapest, for communicating their results prior to publication. We are also indebted to Mr. Manfred Steifa for carrying out calculations using the Cambridge Crystallographic Data File [343], to Mrs. Anna-Liisa Peltola-Hilgenfeld for help in the literature work, to Mr. Ludwig Kolb for drawing several figures, and to Miss Petra Große for typing the manuscript.

10 References

1. Berger, J. et al.: J. Am. Chem. Soc. 73, 5295 (1951)
2. Mueller, P., Rudin, D. O.: Biochem. Biophys. Res. Commun. 26, 398 (1967)
3. Moore, C., Pressman, B. C.: Biochem. Biophys. Res. Commun. 15, 562 (1964)
4. Wipf, H. K., Simon, W.: Helv. Chim. Acta 53, 1732 (1970)
5. Wipf, H. K., Olivier, A., Simon, W.: Helv. Chim. Acta 53, 1605 (1970)
6. Ivanov, V. T.: Ann. N.Y. Acad. Sci. 264, 221 (1975)
7. Gramicidin A has been subject of an (uncompleted) X-ray study: Koeppe, R. E., Hodgson, K. O., Stryer, L.: J. Mol. Biol. 121, 41 (1978); Koeppe, R. E. et al.: Nature 279, 723 (1979)
   A crystallographic study of alamethicin is in progress: Fox, R., private communication
8. Lardy, H.: Fed. Proc. 27, 1278 (1968)
9. Tosteson, D. C.: Fed. Proc. 27, 1269 (1968)
10. Eisenman, G., Ciani, S. M., Szaba, G.: Fed. Proc. 27, 1289 (1968)
11. Harris, E. J. et al.: Arch. Biochem. Biophys. 182, 311 (1977)
12. Pedersen, C. J.: J. Am. Chem. Soc. 89, 7017 (1967)
13. Lehn, J. M.: Acc. Chem. Res. 11, 49 (1978)
14. Vögtle, F., Weber, E.: Angew. Chem. 91, 813 (1979); Angew. Chem. Int. Ed. Engl. 19, 753 (1979)
15. Ovchinnikov, Yu. A., Ivanov, V. T., Shkrob, A. M.: Membrane-active complexones. Amsterdam: Elsevier 1974
16. Burgermeister, W., Winkler-Oswatitsch, R.: Top. Curr. Chem. 69, 91 (1977)

17. Ovchinnikov, Yu. A.: FEBS Lett. *44*, 1 (1974)
18. Pinkerton, M., Steinrauf, L. K., Dawkins, P.: Biochem. Biophys. Res. Commun. *35*, 512 (1969)
19. Neupert-Laves, K., Dobler, M.: Helv. Chim. Acta *58*, 432 (1975)
20. Venkatachalam, C. M.: Biopolymers *6*, 1425 (1968)
21. Steinrauf, L. K., Sabesan, M. N.: Computer simulation studies on the ion-transporting antibiotics. In: Metal-ligand interactions in organic chemistry and biochemistry. Pullman, B., Goldblum, N. (eds.), part 2, pp. 43–57. Dordrecht: D. Reidel 1977
22. Ivanov, V. T. et al.: Biochem. Biophys. Res. Commun. *34*, 803 (1969)
23. Bystrov, V. F. et al.: Eur. J. Biochem. *78*, 63 (1977)
24. Grell, E., Funck, T., Sauter, H.: Eur. J. Biochem. *34*, 415 (1973)
25. Krishna, N. R. et al.: Biophys. J. *24*, 791 (1978)
26. Urry, D. W., Kumar, N. G.: Biochemistry *13*, 1829 (1974)
27. Ohnishi, M. et al.: Biochem. Biophys. Res. Commun. *46*, 312 (1972)
28. Fisher, I. M., Rothschild, K. J., Stanley, H. E.: J. Mol. Biol. *89*, 205 (1974)
29. Mayers, D. F., Urry, D. W.: J. Am. Chem. Soc. *94*, 77 (1972)
30. Duax, W. L. et al.: Science *176*, 911 (1972)
31. Duax, W. L., Hauptman, H.: Acta Cryst. *B28*, 2912 (1972)
32. Karle, I. L.: J. Am. Chem. Soc. *97*, 4379 (1975)
33. Smith, G. D. et al.: J. Am. Chem. Soc. *97*, 7242 (1975)
34. Smith, G. D., Duax, W. L.: Crystallographic studies of valinomycin and A23187. In: Metal-ligand interactions in organic chemistry and biochemistry. Pullmann, B., Goldblum, N. (eds.), part 1, pp. 291–315. Dordrecht: D. Reidel 1977
35. Davies, D. B., Abu Khaled, M.: J. Chem. Soc. Perkin Trans. II, *1976*, 1327
36. Ovchinnikov, Yu. A., Ivanov, V. T.: Tetrahedron *30*, 1871 (1974)
37. Urry, D. W., Kumar, N. G.: Biochemistry *13*, 1829 (1974)
38. Patel, D. J., Tonelli, A. E.: Biochemistry *12*, 486 (1973)
39. Patel, D. J.: Biochemistry *12*, 496 (1973)
40. Glickson, J. D. et al.: Biochemistry *15*, 5721 (1976)
41. Servis, K. L., Patel, D. J.: Tetrahedron *31*, 1359 (1975)
42. Rothschild, K. J. et al.: J. Am. Chem. Soc. *99*, 2032 (1977)
43. Feigenson, G. W., Meers, P. R.: Nature *283*, 313 (1980)
44. Grell, E., Funck, T.: J. Supramol. Struct. *1*, 307 (1973)
45. Ivanov, V. T. et al.: Biochem. Biophys. Res. Commun. *34*, 803 (1969)
46. Maigret, B., Pullman, B.: Theoret. Chim. Acta *37*, 17 (1975)
47. Pletnev, V. Z. et al.: Biopolymers *18*, 2145 (1979)
48. Pletnev, V. Z. et al.: Biopolymers *19*, 1517 (1980)
49. Rothschild, K. J. et al.: Science *182*, 384 (1973)
50. Asher, I. M. et al.: J. Am. Chem. Soc. *99*, 2024 (1977)
51. Shemyakin, M. M. et al.: Biochem. Biophys. Res. Commun. *29*, 834 (1967)
52. Hamill, R. L. et al.: Tetrahedron Lett. *1969*, 4255
53. Plattner, P., Nager, U.: Experientia *3*, 325 (1947)
54. Ovchinnikov, Yu. A. et al.: Int. J. Peptide Protein Res. *6*, 465 (1974)
55. Ovchinnikov, Yu. A. et al.: Biochem. Biophys. Res. Commun. *37*, 668 (1969)
56. Dobler, M., Dunitz, J. D., Krajewski, J.: J. Mol. Biol. *42*, 603 (1969)
57. Vainshtein, B. K. et al.: 4th Eur. Cryst. Meet., Oxford 1977
58. Estrada-O., S., Gomez-Louero, C., Montal, M.: Bioenergetics *3*, 417 (1972)
59. Roeske, R. W. et al.: Biochem. Biophys. Res. Commun. *57*, 554 (1974)
60. Yafuso, M. et al.: Fed. Proc. Abstracts *33*, 1258 (1974)
61. Braden, B. et al.: J. Am. Chem. Soc. *102*, 2704 (1980)
62. Prince, R. C., Crofts, A. R., Steinrauf, L. K.: Biochem. Biophys. Res. Commun. *59*, 697 (1974)
63. Hamilton, J. A., Steinrauf, L. K., Braden, B.: Biochem. Biophys. Res. Commun. *64*, 151 (1975)
64. Geddes, A. J., Akriegg, D.: 4th Eur. Cryst. Meet., Oxford 1977
65. Geddes, A. J., Akrigg, D.: Acta Cryst. *B32*, 3164 (1976)
66. Popov, E. M. et al.: Khim. Priv. Soedin. *5*, 616 (1970)

67. Hamilton, J. A. et al.: Biochem. Biophys. Res. Commun. *80*, 949 (1978)
68. Zhukhlistova, N. E., Tishchenko, G. N.: Kristallografiya, submitted (1981)
69. Karle, I. L. et al.: Proc. Natl. Acad. Sci. USA *70*, 1836 (1973)
70. Karle, I. L.: J. Am. Chem. Soc. *96*, 4000 (1974)
71. Smirnova, V. I., Tishchenko, G. N., Vainshtein, B. K.: Dokl. Akad. Nauk. SSSR, submitted (1981)
72. Karaulov, A. I., Tishchenko, G. N.: Cryst. Struct. Commun., submitted (1981)
73. Tishchenko, G. N. et al.: 6th Eur. Cryst. Meet., Barcelona 1980
74. Tishchenko, G. N., Karaulov, A. I., Karimov, Z.: Cryst. Struct. Commun., submitted (1981)
75. Dunitz, J. D., Dobler, M.: Structural studies of ionophores and their ion-complexes. In: Biological aspects of inorganic chemistry. Addison, A. W. et al. (eds.), pp. 113–140. New York: J. Wiley 1977
76. Shishova, T. G., Simonov, V. I.: Kristallografiya *22*, 515 (1977)
77. Karle, I. L. et al.: Proc. Natl. Acad. Sci. USA *76*, 1532 (1979)
78. Karle, I. L. et al.: Proc. Natl. Acad. Sci. USA *73*, 1782 (1976)
79. Karle, I. L.: J. Am. Chem. Soc. *99*, 5152 (1977)
80. Karle, I. L., Duesler, E.: Proc. Natl. Acad. Sci. USA *74*, 2602 (1977)
81. Dobler, M.: Helv. Chim. Acta *55*, 1371 (1972)
82. Nawata, Y., Sakamaki, T., Iitaka, Y.: Acta Cryst. *B30*, 1047 (1974)
83. Nawata, Y. et al.: Chem. Lett. *1980*, 315
84. Kilbourn, B. T. et al.: J. Mol. Biol. *30*, 559 (1967)
85. Dobler, M., Dunitz, J. D., Kilbourn, B. T.: Helv. Chim. Acta *52*, 2573 (1969)
86. Dobler, M., Phizackerley, R. P.: Helv. Chim. Acta *57*, 664 (1974)
87. Sakamaki, T., Iitaka, Y., Nawata, Y.: Acta Cryst. *B32*, 768 (1976)
88. Iitaka, Y., Sakamaki, T., Nawata, Y.: Chem. Lett. *1972*, 1225
89. Sakamaki, T., Iitaka, Y., Nawata, Y.: Acta Cryst. *B33*, 52 (1977)
90. Neupert-Laves, K., Dobler, M.: Helv. Chim. Acta *59*, 614 (1976)
91. Nawata, Y., Sakamaki, T., Iitaka, Y.: Chem. Lett. *1975*, 151
92. Nawata, Y., Sakamaki, T., Iitaka, Y.: Acta Cryst. *B33*, 1201 (1977)
93. Pioda, L. A. R. et al.: Helv. Chim. Acta *50*, 1373 (1967)
94. Morf, W. E., Simon, W.: Helv. Chim. Acta *54*, 2683 (1971)
95. Szabo, G., Eisenman, G., Ciani, S.: J. Membr. Biol. *1*, 346 (1969)
96. Eisenman, G., Krasne, S., Ciani, S.: Ann. N.Y. Acad. Sci. *264*, 34 (1975)
97. Kyogoku, Y. et al.: Biopolymers *14*, 1049 (1975)
98. Anteunis, M. J. O., De Bruyn, A.: Bull. Soc. Chim. Belg. *86*, 445 (1977)
99. Phillies, G. D. J., Asher, I. M., Stanley, H. E.: Biopolymers *14*, 2311 (1975)
100. Prestegard, J. H., Chan, S. I.: Biochemistry *8*, 3921 (1969)
101. Prestegard, J. H., Chan, S. I.: J. Am. Chem. Soc. *92*, 4440 (1970)
102. Ivanov, V. T. et al.: FEBS Lett. *30*, 199 (1973)
103. Asher, I. M., Phillies, G. D. J., Stanley, H. E.: Biochem. Biophys. Res. Commun. *61*, 1356 (1974)
104. Asher, I. M. et al.: Biopolymers *16*, 157 (1977)
105. Dunitz, J. D. et al.: Helv. Chim. Acta *54*, 1709 (1971)
106. Marsh, W., Dunitz, J. D., White, D. N. J.: Helv. Chim. Acta *57*, 10 (1974)
107. Gertenbach, P. G., Popov, A. I.: J. Am. Chem. Soc. *97*, 4738 (1975)
108. Ward, D. L. et al.: Acta Cryst. *B34*, 110 (1978)
109. Westley, J. W.: Ann. Rep. Med. Chem. *10*, 246 (1975)
110. Westley, J. W.: Adv. Appl. Microbiol. *22*, 177 (1977)
111. Ebata, E. et al.: J. Antibiot. *28*, 118 (1975)
112. Liu, W.-C. et al.: J. Antibiot. *31*, 815 (1978)
113. Liu, C., Hermann, T. E.: J. Biol. Chem. *253*, 5892 (1978)
114. Liu, C. et al.: J. Antibiot. *32*, 95 (1979)
115. Lutz, W. K., Wipf, H.-K., Simon, W.: Helv. Chim. Acta *53*, 1741 (1970)
116. Pressman, B. C., Haynes, D. H.: Ionophore agents as mobile ion carriers. In: The molecular basis of membrane function. Tosteson, D. C. (ed.), pp. 221–246. Englewood Cliffs, N. Y.: Prentice Hall 1969
117. Pressman, B. C.: Ann. Rev. Biochem. *45*, 4925 (1976)

118. Agtarap, A. et al.: J. Am. Chem. Soc. *89*, 5737 (1967)
119. Pinkerton, M., Steinrauf, L. K.: J. Mol. Biol. *49*, 533 (1970)
120. Duax, W. L., Smith, G. D., Strong, P. D.: J. Am. Chem. Soc. *102*, 6725 (1980)
121. Alléaume, M., Hickel, D.: J.C.S. Chem. Commun. *1970*, 1422
122. Alléaume, M., Hickel, D.: J.C.S. Chem. Commun. *1972*, 175
123. Lutz, W. K., Winkler, F. K., Dunitz, J. D.: Helv. Chim. Acta *54*, 1103 (1971)
124. Haynes, D. H., Pressman, B. C., Kowalsky, A.: Biochemistry *10*, 852 (1971)
125. Anteunis, M. J. O.: Bull. Soc. Chim. Belg. *86*, 367 (1977)
126. Anteunis, M. J. O., Rodios, N. A.: Biorg. Chem. *7*, 47 (1978)
127. Briggs, R. W., Hinton, J. F.: Biochemistry *17*, 5576 (1978)
128. Painter, G., Pressman, B. C.: Biochem. Biophys. Res. Commun. *91*, 1117 (1979)
129. Young, S. P., Comperts, B. D.: Biochim. Biophys. Acta *469*, 281 (1977)
130. Westley, J. W., Evans, R. H., Blount, J. F.: J. Am. Chem. Soc. *99*, 6057 (1977)
131. Johnson, S. M. et al.: J.C.S. Chem. Commun. *1970*, 72
132. Johnson, S. M. et al.: J. Am. Chem. Soc. *92*, 4428 (1970)
133. Maier, C. A., Paul, I. C.: J.C.S. Chem. Commun. *1971*, 181
134. Schmidt, P. G., Wang, A. H. J., Paul, I. C.: J. Am. Chem. Soc. *96*, 6189 (1974)
135. Bissell, E. C., Paul, I. C.: J.C.S. Chem. Commun. *1972*, 967
136. Patel, D. J., Shen, C.: Proc. Natl. Acad. Sci. USA *73*, 1786 (1976)
137. Shen, C., Patel, D. J.: Proc. Natl. Acad. Sci. USA *73*, 4277 (1976)
138. Chiang, C. C., Paul, I. C.: Science *196*, 1441 (1977)
139. Friedman, J. M. et al.: J. Chem. Soc. Perkin Trans. II *1979*, 835
140. Smith, G. D., Duax, W. L., Fortier, S.: J. Am. Chem. Soc. *100*, 6725 (1978)
141. Chaney, M. O. et al.: J. Am. Chem. Soc. *96*, 1932 (1974)
142. Chaney, M. O., Jones, N. D., Debono, M.: J. Antibiot. *29*, 424 (1976)
143. Smith, G. D., Duax, W. L.: J. Am. Chem. Soc. *98*, 1578 (1976)
144. Perlman, R. L., Cossi, A. F., Role, L. W.: J. Pharmacol. Exp. Ther. *213*, 241 (1980)
145. Toeplitz, B. K. et al.: J. Am. Chem. Soc. *101*, 3344 (1979)
146. Barrans, Y., Alléaume, M.: Acta Cryst. *B36*, 936 (1980)
147. Steinrauf, L. K., Pinkerton, M., Chamberlin, J. W.: Biochem. Biophys. Res. Commun. *33*, 29 (1968)
148. Shiro, M., Koyama, H.: J. Chem. Soc. (B) *1970*, 243
149. Kubota, T. et al.: J.C.S. Chem. Commun. *1968*, 1541
150. Geddes, A. J.: Biochem. Biophys. Res. Commun. *60*, 1245 (1974)
151. Riche, C., Pascard-Billy, C.: J.C.S. Chem. Commun. *1975*, 951
152. Yamazaki, K., Abe, K., Sano, M.: J. Antibiot. *29*, 91 (1976)
153. Otake, N., Koenuma, M.: Tetrahedron. Lett. *1975*, 4147
154. Otake, N. et al.: J. Chem. Soc. Perkin Trans. II *1977*, 494
155. Blount, J. F., Westley, J. W.: J.C.S. Chem. Commun. *1971*, 927
156. Blount, J. F., Westley, J. W.: J.C.S. Chem. Commun. *1975*, 533
157. Alléaume, M. et al.: J.C.S. Chem. Commun. *1975*, 411
158. Czerwinski, E. W., Steinrauf, L. K.: Biochem. Biophys. Res. Commun. *45*, 1284 (1971)
159. Blount, J. F. et al.: J.C.S. Chem. Commun. *1975*, 853
160. Koyama, H., Utsumi-Oda, K.: J. Chem. Soc. Perkin Trans. II *1977*, 1531
161. Jones, N. D. et al.: J. Am. Chem. Soc. *95*, 3399 (1973)
162. Otake, N. et al.: J.C.S. Chem. Commun. *1977*, 590
163. Nakayama, H. et al.: J. Chem. Soc. Perkin Trans. II *1979*, 293
164. Shiro, M. et al.: J.C.S. Chem. Commun. *1978*, 682
165. Otake, N. et al.: J.C.S. Chem. Commun. *1978*, 875
166. Otake, N. et al.: J.C.S. Chem. Commun. *1975*, 92
167. Koenuma, M., Kinashi, H., Otake, N.: Acta Cryst. *B32*, 1267 (1976)
168. Alléaume, M.: 2nd. Eur. Cryst. Meet., Keszthely, Hungary, 1974
169. Kinashi, H. et al.: Tetrahedron. Lett. *1973*, 4955
170. Kinashi, H. et al.: Acta Cryst. *B31*, 2411 (1975)
171. Petcher, T. J., Weber, H. P.: J.C.S. Chem. Commun. *1974*, 697
172. Smith, G. D., Strong, P. D., Duax, W. L.: Acta Cryst. *B34*, 3436 (1978)
173. Westley, J. W. et al.: J. Antibiot. *27*, 597 (1974)

174. Westley, J. W. et al.: J. Am. Chem. Soc. *100*, 6784 (1978)
175. Westley, J. W. et al.: J. Antibiot. *32*, 100 (1979)
176. Van Remoortere, F. P., Boer, F. P.: Inorg. Chem. *13*, 2071 (1974)
177. Boer, F. P. et al.: Inorg. Chem. *13*, 2826 (1974)
178. Neuman, M. A. et al.: Inorg. Chem. *14*, 734 (1975)
179. North, P. P. et al.: Acta Cryst. *B32*, 370 (1976)
180. Van Remoortere, F. P., Boer, F. P., Steiner, E. C.: Acta Cryst. *B31*, 1420 (1975)
181. Armağan, N.: Acta Cryst. *B33*, 2281 (1977)
182. Feneau-Dupont, J. et al.: Acta Cryst. *B35*, 1217 (1979)
183. Arte, E. et al.: Acta Cryst. *B35*, 1215 (1979)
184. Bush, M. A., Truter, M. R.: J. Chem. Soc. Perkin Trans. II, *1972*, 341
185. Mallinson, P. R., Truter, M. R.: J. Chem. Soc. Perkin Trans. II, *1972*, 1818
186. Owen, J. D.: J. Chem. Soc. Dalton Trans. *1978*, 1418
187. Owen, J. D., Wingfield, J. N.: J.C.S. Chem. Commun. *1976*, 318
188. Cradwick, P. D., Poonia, N. S.: Acta Cryst. *B33*, 197 (1977)
189. Bhagwat, V. W., Manohar, H., Poonia, N. S.: 11th Internat. Congr. Cryst., Warsaw, 1978; J. Inorg. Nucl. Chem. Lett. *16*, 373 (1980)
190. Dobler, M., Dunitz, J. D., Seiler, P.: Acta Cryst. *B30*, 2741 (1974)
191. Sheldrick, W. S. et al.: Angew. Chem. *91*, 998 (1979); Angew. Chem. Int. Ed. Engl. *18*, 934 (1979)
192. Seiler, P., Dobler, M., Dunitz, J. D.: Acta Cryst. *B30*, 2744 (1974)
193. Riche, C. et al.: J.C.S. Chem. Commun. *1977*, 183
194. Groth, P.: Acta Chem. Scand. *25*, 3189 (1971)
195. Nagano, O.: Acta Cryst. *B35*, 465 (1979)
196. Nagano, O., Sasaki, Y.: Acta Cryst. *B35*, 2387 (1979)
197. Dobler, M., Phizackerley, R. P.: Acta Cryst. *B30*, 2746 (1974)
198. Dobler, M., Phizackerley, R. P.: Acta Cryst. *B30*, 2748 (1974)
199. Dunitz, J. D., Seiler, P.: Acta Cryst. *B30*, 2750 (1974)
200. Vance, T. B. et al.: Acta Cryst. *B36*, 153 (1980)
201. Vance, T. B. et al.: Acta Cryst. *B36*, 150 (1980)
202. Bombieri, G., De Paoli, G., Immirzi, A.: J. Inorg. Nucl. Chem. *40*, 1889 (1978)
203. Bombieri, G. et al.: Inorg. Chim. Acta *18*, L23 (1976)
204. Hašek, J., Huml, K.: Acta Cryst. *B34*, 1812 (1978)
205. Hughes, D. L., Mortimer, C. L., Truter, M. R.: Inorg. Chim. Acta *29*, 43 (1978)
206. Hlavatá, D., Hašek, J., Huml, K.: Acta Cryst. *B34*, 416 (1978)
207. Hašek, J., Hlavatá, D., Huml, K.: Acta Cryst. *B33*, 3372 (1977)
208. Bush, M. A., Truter, M. R.: J. Chem. Soc. (B) *1971*, 1440
209. Bright, D., Truter, M. R.: Nature *225*, 176 (1970)
210. Bright, D., Truter, M. R.: J. Chem. Soc. (B) *1970*, 1544
211. Myskiv, M. G. et al.: 11th Internat. Congr. Cryst., Warsaw 1978
212. Hilgenfeld, R., Saenger, W.: 6th Eur. Cryst. Meet., Barcelona 1980; Angew. Chem., in the press
213. Ciampolini, M. et al.: J. Chem. Soc. Dalton Trans. *1979*, 1983
214. Mallinson, P. R.: J. Chem. Soc. Perkin Trans. II, *1975*, 261
215. Layton, A. J. et al.: J.C.S. Chem. Commun. *1973*, 694
216. Fenton, D. E., Mercer, M., Truter, M. R.: Biochem. Biophys. Res. Commun. *48*, 10 (1972)
217. Mercer, M., Truter, M. R.: J. Chem. Soc. Dalton Trans. *1973*, 2215
218. Dalley, N. K. et al.: J.C.S. Chem. Commun. *1972*, 90
219. Harman, M. E. et al.: J.C.S. Chem. Commun. *1976*, 396
220. De Villardi, G. C. et al.: J.C.S. Chem. Commun. *1978*, 90
221. Ward, D. L., Brown, H. S., Sousa, L. R.: Acta Cryst. *B33*, 3537 (1977)
222. Czugler, M.: private communication
223. Hughes, D. L.: J. Chem. Soc. Dalton Trans. *1975*, 2374
224. Fenton, D. E. et al.: J.C.S. Chem. Commun. *1972*, 66
225. Mercer, M., Truter, M. R.: J. Chem. Soc. Dalton Trans. *1973*, 2469
226. Hughes, D. L., Mortimer, C. L., Truter, M. R.: Acta Cryst. *B34*, 800 (1978)
227. Hughes, D. L., Wingfield, J. N.: J.C.S. Chem. Commun. *1977*, 804
228. Owen, J. D., Truter, M. R.: J. Chem. Soc. Dalton Trans. *1979*, 1831

229. Bush, M. A., Truter, M. R.: J. Chem. Soc. Perkin Trans. II *1972*, 345
230. Hašek, J., Hlavatá, D., Huml, K.: Acta Cryst. *B36*, 1782 (1980)
231. Hašek, J., Huml, K., Hlavatá, D.: Acta Cryst. *B35*, 330 (1979)
232. M. Czugler and E. Weber: J.C.S. Chem. Commun. *1981*, 472
233. Moras, D. et al.: Bull. Soc. Chim. France *1972*, 551
234. Metz, B., Weiss, R.: Acta Cryst. *B29*, 1088 (1973)
235. Malmsten, L.-Å.: Acta Cryst. *B35*, 1702 (1979)
236. Herceg, M., Weiss, R.: Acta Cryst. *B29*, 542 (1973)
237. Herceg, M., Weiss, R.: Rev. Chim. Min. *10*, 509 (1973)
238. Groth, P.: Acta Chem. Scand. *A32*, 279 (1978)
239. Hanson, I. R.: Acta Cryst. *B34*, 1026 (1978)
240. Dunitz, J. D., Seiler, P.: Acta Cryst. *B30*, 2739 (1974)
241. Maverick, E. et al.: Acta Cryst. *B36*, 615 (1980)
242. Truter, M. R.: Effects of cations of groups IA and IIA on crown ethers. In: Metal-ligand interactions in organic chemistry and biochemistry. Pullman, B., Goldblum, N. (eds.), part. 1, pp. 317–335. Dordrecht: D. Reidel 1977
243. Dalley, N. K. et al.: J.C.S. Chem. Commun. *1975*, 43
244. Dalley, N. K.: Structural studies of synthetic macrocyclic molecules and their cation complexes. In: Synthetic multidentate macrocyclic compounds. Izatt, R. M., Christensen, J. J. (eds.), pp. 207–243. New York: Academic Press 1978
245. Mallinson, P. R.: J. Chem. Soc. Perkin Trans. II *1975*, 266
246. Dalley, N. K., Larson, S. B.: Acta Cryst. *B35*, 1901 (1979)
247. Czugler, M.: private communication
248. Goldberg, I.: Acta Cryst. *B34*, 2224 (1978)
249. Goldberg, I.: Acta Cryst. *B32*, 41 (1976)
250. Owen, J. D., Nowell, I. W.: Acta Cryst. *B34*, 2354 (1978)
251. Dalley, N. K., Larson, S. B.: Acta Cryst. *B35*, 2428 (1979)
252. Fronczek, F. R., Nayak, A., Newkome, G. R.: Acta Cryst. *B35*, 775 (1979)
253. Hanson, I. R., Hughes, D. L., Truter, M. R.: J. Chem. Soc. Perkin Trans. II *1976*, 972
254. Herceg, M., Weiss, R.: Bull. Soc. Chim. Fr. *1972*, 549
255. Dalley, N. K. et al.: J.C.S. Chem. Commun. *1975*, 84
256. Tables of interatomic distances and configurations in molecules and ions, London: The Chemical Society 1960
257. Dale, J.: Tetrahedron Lett. *30*, 1683 (1974)
258. Dale, J., Kristiansen, P. O.: Acta Chem. Scand. *26*, 1471 (1972)
259. Bovill, M. J. et al.: J. Chem. Soc. Perkin Trans. II *1980*, 1529
260. Frensdorff, H. K.: J. Am. Chem. Soc. *93*, 600 (1971)
261. Poonia, N. S., Bajaj, A. V.: Chem. Rev. *79*, 389 (1979)
262. Poonia, N. S. et al.: Ind. J. Chem. *19A*, 37 (1980)
263. Live, D., Chan, S. I.: J. Am. Chem. Soc. *98*, 3769 (1976)
264. Mei, E., Dye, J. L., Popov, A. I.: J. Am. Chem. Soc. *98*, 1619 (1976)
265. Popov, A. I.: Multinuclear NMR studies of crown and cryptand complexes. In: Stereodynamics of molecular systems. Sarma, R. H. (ed.), pp. 197–207. New York: Pergamon Press 1979
266. Krane, J., Dale, J., Daasvatn, K.: Acta Chem. Scand. *B34*, 59 (1980)
267. Parsons, D. G., Truter, M. R., Wingfield, J. N.: Inorg. Chim. Acta *14*, 45 (1975)
268. Pedersen, C. J.: J. Org. Chem. *36*, 1690 (1971)
269. Moras, D., Weiss, R.: Acta Cryst. *B29*, 400 (1973)
270. Mathieu, F. et al.: J. Am. Chem. Soc. *100*, 4412 (1978)
271. Mathieu, F., Weiss, R.: J.C.S. Chem. Commun. *1973*, 816
272. Moras, D., Weiss, R.: Acta Cryst. *B29*, 396 (1973)
273. Teller, R. G. et al.: J. Am. Chem. Soc. *99*, 1104 (1977)
274. Ginsburg, R. E. et al.: J. Am. Chem. Soc. *101*, 6550 (1979)
275. Tehan, F. J., Barnett, B. L., Dye, J. L.: J. Am. Chem. Soc. *96*, 7203 (1974)
276. Adolphson, D. G., Corbett, J. D., Merryman, D. J.: J. Am. Chem. Soc. *98*, 7234 (1976)
277. Corbett, J. D., Edwards, P. A.: J. Am. Chem. Soc. *99*, 3313 (1977)
278. Moras, D., Metz, B., Weiss, R.: Acta Cryst. *B29*, 383 (1973)
279. Moras, D., Metz, B., Weiss, R.: Acta Cryst. *B29*, 388 (1973)

280. Metz, B., Moras, D., Weiss, R.: Acta Cryst. *B29*, 1377 (1973)
281. Metz, B., Moras, D., Weiss, R.: Acta Cryst. *B29*, 1382 (1973)
282. Moras, D., Weiss, R.: Acta Cryst. *B29*, 1059 (1973)
283. Metz, B., Weiss, R.: Inorg. Chem. *13*, 2094 (1974)
284. Hart, F. A. et al.: J.C.S. Chem. Commun. *1978*, 549
285. Ciampolini, M., Dapporto, P., Nardi, N.: J.C.S. Chem. Commun. *1978*, 788
286. Ciampolini, M., Dapporto, P., Nardi, N.: J. Chem. Soc. Dalton Trans. *1979*, 974
287. Burns, J. H.: Inorg. Chem. *18*, 3044 (1979)
288. Metz, B., Moras, D., Weiss, R.: Acta Cryst. *B29*, 1388 (1973)
289. Lehn, J. M.: Molecular receptors, carriers, and catalysts: Design, scope, and prosepcts. In: Bioenergetics and thermodynamics: Model Systems. Braibanti, A. (ed.), pp. 455–461. Dordrecht: D. Reidel 1980
290. Lehn, J. M., Sauvage, J. P.: J. Am. Chem. Soc. *97*, 6700 (1975)
291. Metz, B., Moras, D., Weiss, R.: J. Chem. Soc. Perkin Trans. II *1976*, 423
292. Metz, B., Weiss, R.: Nouv. J. Chim. *2*, 615 (1978)
293. Metz, B., Moras, D., Weiss, R.: 2nd Eur. Cryst. Meet., Keszthely, Hungary, 1974
294. Mei, E., Popov, A. I., Dye, J. L.: J. Am. Chem. Soc. *99*, 6532 (1977)
295. Mei, E. et al.: J. Solution Chem. *6*, 771 (1977)
296. Kauffmann, E., Lehn, J. M., Sauvage, J. P.: Helv. Chim. Acta *59*, 1099 (1976)
297. Saenger, W., Reddy, B. S.: Acta Cryst. *B35*, 56 (1979)
298. Saenger, W. et al.: X-ray structure of a synthetic, non-cyclic, chiral polyether complex as analog of nigericin antibiotics. In: Metal-ligand interactions in organic chemistry and biochemistry. Pullman, B., Goldblum, N. (eds.), part 1, pp. 363–374. Dordrecht: D. Reidel 1977
299. Saenger, W., Brand, H.: Acta Cryst. *B35*, 838 (1979)
300. Hilgenfeld, R., Saenger, W.: unpublished results
301. Weber, G., Saenger, W.: Acta Cryst. *B35*, 1346 (1979)
302. Weber, G. et al.: Angew. Chem. *91*, 234 (1979); Angew. Chem. Int. Ed. Engl. *18*, 226 (1979)
303. Weber, G., Saenger, W.: Acta Cryst. *B35*, 3093 (1979)
304. Weber, G., Saenger, W.: Angew. Chem. *91*, 237 (1979); Angew. Chem. Int. Ed. Engl. *18*, 227 (1979)
305. Weber, G., Saenger, W.: Acta Cryst. *B36*, 61 (1980)
306. Suh, I.-H., Weber, G., Saenger, W.: Acta Cryst. *B34*, 2752 (1978)
307. Suh, I.-H., Weber, G., Saenger, W.: Acta Cryst. *B36*, 946 (1980)
308. Chacko, K. K., Saenger, W.: Z. Naturforsch. *35b*, 1533 (1980)
309. Czugler, M.: private communication
310. Suh, I.-H. et al.: Z. Naturforsch. *35b*, 352 (1980)
311. Chacko, K. K., Saenger, W.: Z. Naturforsch. *36b*, 102 (1981)
312. Hughes, D. L. et al.: Inorg. Chim. Acta *21*, L23 (1977)
313. Hughes, D. L., Mortimer, C. L., Truter, M. R.: Inorg. Chim. Acta *28*, 83 (1978)
314. Hughes, D. L., Wingfield, J. N.: J.C.S. Chem. Commun. *1978*, 1001
315. Singh, T. P., Reinhardt, R., Poonia, N. S.: Inorg. Nucl. Chem. Lett. *16*, 293 (1980)
316. Ohmoto, H. et al.: Bull. Chem. Soc. Jpn. *52*, 1209 (1979)
317. Hilgenfeld, R., Saenger, W.: Z. Anal. Chem. *304*, 277 (1980)
318. Weber, G.: Acta Cryst. *B36*, 2779 (1980)
319. Iwamoto, R.: Bull. Chem. Soc. Jpn. *46*, 1114 (1973)
320. Iwamoto, R.: Bull. Chem. Soc. Jpn. *46*, 1123 (1973)
321. Iwamoto, R., Wakano, H.: J. Am. Chem. Soc. *98*, 3764 (1976)
322. Saenger, W., Suh, I.-H.: unpublished results
323. Weber, G., Sheldrick, G. M.: Inorg. Chim. Acta *45*, L35 (1980)
324. Izatt, R. M., Eatough, D. J., Christensen, J. J.: Struc. Bonding *16*, 161 (1973)
325. Cabbiness, D. K., Margerum, D. W.: J. Am. Chem. Soc. *91*, 6540 (1969)
326. Kodama, M., Kimura, E.: Bull. Chem. Soc. Jpn. *49*, 2465 (1976)
327. Fabbrizzi, L., Paoletti, P., Lever, A. B. P.: Inorg. Chem. *15*, 1502 (1976)
328. Hancock, R. D., McDougall, G. J.: J. Am. Chem. Soc. *102*, 6551 (1980)
329. Vögtle, F., Sieger, H.: Angew. Chem. *89*, 410 (1977); Angew. Chem. Int. Ed. Engl. *16*, 396 (1977)
330. Saenger, W., Suh, I.-H., Weber, G.: Isr. J. Chem. *18*, 253 (1979)
331. Tümmler, B. et al.: J. Am. Chem. Soc. *101*, 2588 (1979)

332. Hughes, D. L., Truter, M. R.: J. Chem. Soc. Dalton Trans. *1979*, 520
333. Yamazaki, N. et al.: Tetrahedron Lett. *1978*, 2429
334. Sieger, H., Vögtle, F.: Liebigs Ann. Chem. *1980*, 425
335. Grandjean, J. et al.: Angew. Chem. *90*, 902 (1978); Angew. Chem. Int. Ed. Engl. *17*, 856 (1978)
336. Sieger, H., Vögtle, F.: Angew. Chem. *90*, 212 (1978); Angew. Chem. Int. Ed. Engl. *17*, 198 (1978)
337. Hilgenfeld, R., Saenger, W.: unpublished results
338. Vögtle, F. et al.: Angew. Chem. *89*, 564 (1977); Angew. Chem. Int. Ed. Engl. *16*, 548 (1977)
339. Weber, G., Sheldrick, G. M.: Acta Cryst. *B36*, 1978 (1980)
340. Goldberg, I.: Acta Cryst. *B31*, 754 (1975)
341. Hilgenfeld, R., Saenger, W.: Z. Naturforsch. *36b*, 242 (1981)
342. Dunitz, J. D., Ha, T. K.: J.C.S. Chem. Commun. *1972*, 568
343. Allen, F. H. et al.: Acta Cryst. *B35*, 2331 (1979)
344. Christensen, J. J.: Transport of metal ions by liquid membranes containing macrocyclic carriers. In: Bioenergetics and thermodynamics: Model systems. Braibanti, A. (ed.), pp. 111 to 126. Dordrecht: D. Reidel 1980
345. Vögtle, F., Sieger, H., Müller, W. M.: Top. Curr. Chem. *98*, 107 (1981)
346. Cram, D. J., Trueblood, K. N.: Top. Curr. Chem. *98*, in the press

Dynamic Aspects of Ionophore Mediated Membrane Transport

George R. Painter and Berton C. Pressman

Department of Pharmacology, University of Miami School of Medicine, Miami, Florida 33101, USA

Table of Contents

1 Historical Introduction 84

2 Ionophore Structure 85
  2.1 General Architectural Features 85
  2.2 Structures of Representative Ionophores 86

3 Ionophore-Mediated Ion Transport in Membranes 88
  3.1 Neutral Ionophores 88
  3.2 Carboxylic Ionophores 89

4 Conformational Dynamics of Ionophores 90
  4.1 Macrocyclic Neutral Ionophores in Bulk Solvents 90
  4.2 Carboxylic Ionophores in Bulk Solvents 91
  4.3 Carboxylic Ionophore Conformational Dynamics During Transport . . 99

5 Specific Membrane-Ionophore Interactions 100
  5.1 Membrane Composition and Ionophore Activity 100
  5.2 Specific Carboxylic Ionophore-Phospholipid Interactions 101

6 Biological Properties of Ionophores 102
  6.1 Cardiovascular Properties of Carboxylic Ionophores 102
  6.2 Mechanism of Action of Carboxylic Ionophores 104
  6.3 Transport of Na^+, K^+ and H^+ across RBC Membranes 105

7 Conclusion and Future Prospects 107

8 References 108

1 Historical Introduction

The first indication of the unique biological and physical properties of the ionophores was the serendipitous discovery by McMurray and Beggs that valinomycin is a potent uncoupler of mitochondrial oxidative phosphorylation [1]. Pressman later observed that valinomycin released more protons from mitochondria than could be accounted for on the basis of ATP hydrolysis [2] (Fig. 1). Upon adding valinomycin to a medium

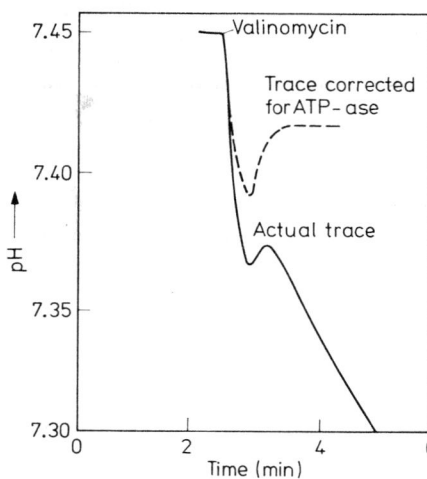

Fig. 1. Recording of the pH response of a mitochondrial system (ATP, MgCl$_2$, KCl, TRIS and sucrose) to valinomycin. The solid trace was obtained experimentally. The dashed curve was derived by correcting the trace for H$^+$ production due to ATP hydrolysis, as determined directly by phosphate analysis

containing mitochondria, ATP and K$^+$, a transient burst of acidification of the medium was observed which spontaneously halted, reversed and resumed. When the curve obtained by monitoring the medium with a pH electrode was corrected for the acid equivalents attributable to ATP hydrolysis, a derived curve (dashed) was obtained representing an energy dependent expulsion of protons from the mitochondria [3]. This phenomenon has been termed the "drainpipe" effect [4]. The true nature of the ionophore mediated mitochondrial process as an energy-dependent cation transport (K$^+$, H$^+$) phenomenon was established by Moore and Pressman using ion selective electrodes [3]. The process could be equally supported by ATP or oxidizable substrate as a source of energy and consisted of the uptake of K$^+$ by mitochondria against a concentration gradient in exchange for H$^+$. When the energy source was cut off with an appropriate inhibitor, the ion gradients relaxed toward their initial state. It was subsequently found that the macrotetralide actins [5], the enniatins [6] and certain of their synthetic analogs also initiate the energy-linked accumulation of alkali ions by mitochondria.

A series of valinomycin analogs in which the ring size was enlarged or in which a single asymmetric center was inverted were found to be much less active than the parent valinomycin [5]. This information together with the extremely low concentrations of valinomycin required ($\sim 10^{-9}$ M), was interpreted as suggesting that valinomycin activated a specific mitochondrial K$^+$ transport receptor.

Graven et al. observed that a second class of antibiotics, exemplified by nigericin, could reverse the changes induced in mitochondria by valinomycin [7]. It was concluded that antibiotics of this type present mitochondria with ion complexes which inhibit the mitochondrial ion pump, thereby permitting accumulated ions to leak back down the existing intra-extramitochondrial concentration gradient (Lardy et al., 1977).

Mueller and Rudin [8] and Lev and Buzhinsky [9] found, however, that valinomycin induced an increase in the ion permeability of black lipid membranes which obviously were free of all complicated biological transducers. An alternative mechanism was proposed to explain the activity of valinomycin in which it was suggested that valinomycin forms an ion conducting channel across the lipid bilayer [8]. This mechanism was rejected in favor of the currently accepted *mobile carrier* mechanism when it was demonstrated that both valinomycin and nigericin could transport ions across bulk phases too thick to accommodate stacked molecular channels [10]. The generic term *ionophore*, i.e. ion carrier, was chosen to describe these compounds by Pressman et al. in order to emphasize the dynamic aspects of the transport mechanism [6,10]. Although this term is widely accepted, Ovchinnikov et al. choose to use the term *complexone* to emphasize the nature of the association between the carrier molecule and the cation [11].

2 Ionophore Structure

2.1 General Architectural Features

In order to function as efficient mobile carriers within natural or artificial membranes, ionophores must possess a distinct set of functional and structural characteristics.

1. The ionophores must possess both polar and non-polar groups. The polar groups function as ligands which replace the solvent molecules in the primary solvation sphere of the complexed ion. In order to compete with the ionic solvation sphere, the ionophore should contain five to eight but not more than twelve liganding polar groups. The most stable ionophore-alkali ion inclusion complexes have a co-ordination number of six, while the larger alkali earth cations prefer a coordination number of eight [12].
2. The ionophore must be able to assume a stable conformation which directs the polar liganding moieties into a central cavity suitable for encaging a cation [13]. High ion complexation selectivities are achieved by locking the coordination sites into a rigid arrangement around the cavity. Rigidity is generally imparted by intramolecular hydrogen bonding, extended spirane systems and substituents which limit free rotation about key bonds. The complexing conformation must also result in the alkyl substituents on the ionophore backbone being directed outward. These alkyl groups must insulate the internal liganding cavity sufficiently to insure the stability of the ion while crossing the apolar membrane interior.
3. Ion complexation-decomplexation reactions must proceed at a sufficiently rapid rate in order for the ionophore to function as an efficient carrier. This is only possible when the ionophore is flexible enough to allow a stepwise rather than a concerted substitution of liganding moieties for solvent molecules [14]. Thus,

in order to be an effective ionophore a compromise between ion affinity and ion exchange kinetics must exist [15,16].

2.2 Structures of Representative Ionophores

Numerous structurally diverse compounds have been found to act as ionophores within biological membranes [11,17]. The general structural features of several representative ionophorous agents are illustrated in Fig. 2. Heteroatoms which have been identified as liganding moieties by X-ray crystallography are filled in.

Valinomycin (Fig. 2A) consists of alternating residues of hydroxyacids and amino acids constituting a cyclic *dodecadepsipeptide* [18]. In space, the peptide backbone undulates three times defining a bracelet 4 Å high and 8 Å in diameter [19,20]. The liganding oxygens, the ester carbonyls, form a three dimensional cage which accommodates K^+ (r = 1.33 Å) much more snugly than Na^+ (r = 0.95 Å) resulting in a $K^+:Na^+$ preference of 10,000 to 1 [10,21]. Over eighty valinomycin analogs have been synthesized [11].

Enniatin B (Fig. 2B) is a cyclic hexadepsipeptide [22,23]. The ring, which is only half the size of that of valinomycin, is too small to fold into the bracelet conformation of valinomycin and forms a staggered plane [24]. The complexing cage is consequently less well defined and the $K^+:Na^+$ selectivity ($K^+:Na^+ \sim 3:1$) is considerably attenuated [11].

The macrolide nactins (Fig. 2C) are constructed of four tetrahydrofuranyl hydroxyacids linked together by ester bonds [25]. The liganding tetrahydrofuranyl oxygens are arranged at the apices of a cubic cage [25A]. Five varient, homologous nactins are known depending on whether the R groups are methyls (nonactin) or are 1–4 ethyls (monactin, dinactin, trinactin, tetranactin). The stereochemistry of the macrolide nactins is particularly interesting in that in all cases, the optical configurations of the consecutive hydroxyacids alternate.

While the aforementioned ionophores are microbial metabolites, the crown polyethers, the depicted prototype of which is dicyclohexyl-18-crown-6 (Fig. 2D) are synthetic macrocyclic ethers. The first crown ether synthesized, dibenzo-18-crown-6, is the first multidentate synthetic macrocycle with the ability to form stable complexes with alkali and alkaline earth compounds [26]. The complexing abilities of this crown ether led to the preparation of many others in rapid succession. Different size crown rings have been synthesized containing benzyl, cyclohexyl and naphthyl moieties [27]. Optically pure dinaphthyl crown ethers have been used to resolve asymmetric amine salts by enantioselective extraction from an aqueous solution into chloroform [28,29].

The ionophores thus far described contain no ionizable functional groups and are collectively classified as *neutral* ionophores [17]. Ion complexation by neutral ionophores results in a complex carrying the net charge of the included ion. The *carboxylic* subclass of ionophores all contain a terminal carboxyl group which is fully ionized at physiological pH [30]. Formation of ion inclusion complexes by carboxylic ionophores gives rise to an electrically neutral zwitterion. This distinction is fundamental for explaining the profound differences in the biological behavior of the ionophore subclasses [17]. Hence we prefer *carboxylic ionophore* to the term

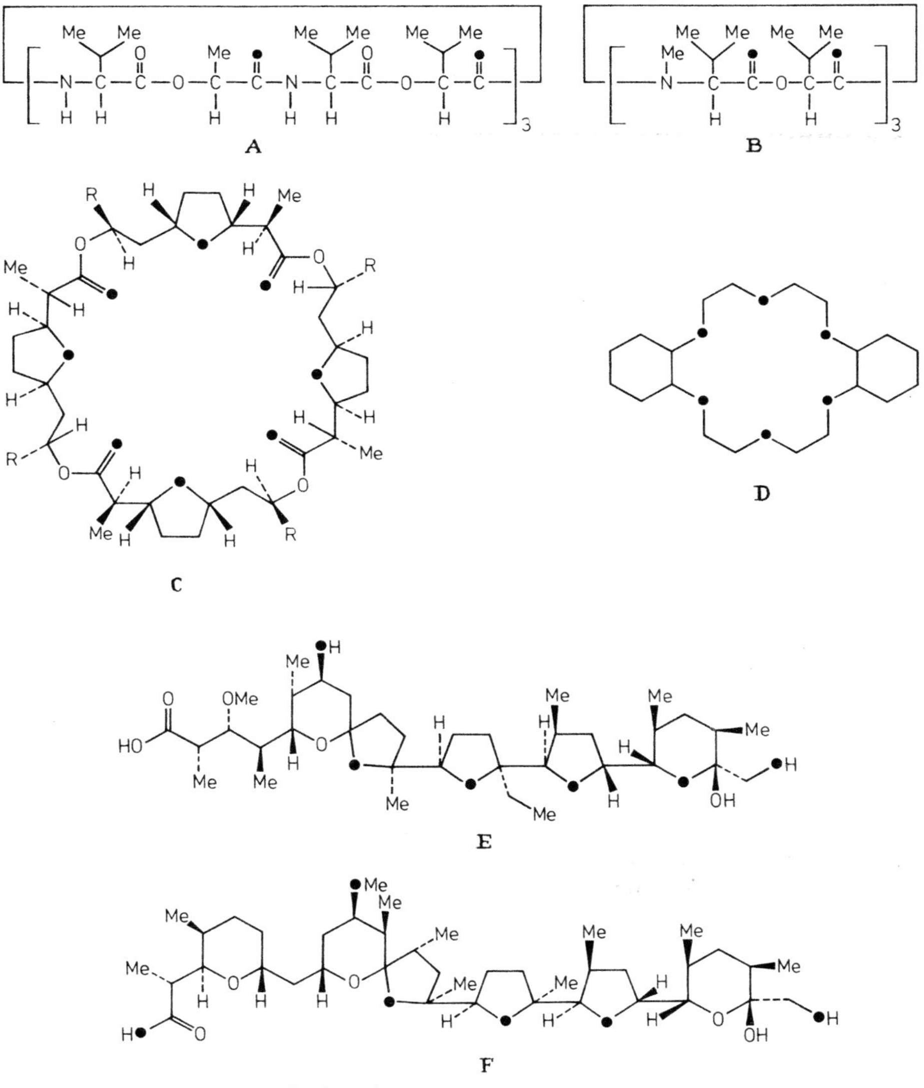

Fig. 2. Structure of representative ionophores

polyether antibiotic used by Westley [31]. The latter term, furthermore, leads to functional ambiguity with the macrolide nactins and crown ethers and presumes that all ionophores have significant antibiotic activity.

The naturally occurring carboxylic ionophores, typified by monensin, lack the structural redundancy of the neutral ionophores [32, 33]. Although monensin (Fig. 2E) consists of a formally linear array of tetrahydrofuranyl and tetrahydropyranyl rings, the molecular chirality arising from the rings and the asymmetric carbons on the ionophore backbone favors the molecule assuming a quasi-cyclic configuration. Additional stabilization of the resulting macrocycle is conferred by head-to-tail hydrogen bonding. In addition to the liganding ether oxygens, monensin has two liganding hydroxyl oxygens.

Rings A through E of nigericin (Fig. 2F) closely resemble rings A through E of monensin [34, 35]. However, an additional tetrahydropyranyl ring (ring F) thrusts the carboxyl group of nigericin rigidly into the cation complexation cavity. Thus, in addition to the induced dipole-ion bonds previously described, nigericin complexes feature a true ionic bond. Despite major similarities in structure, nigericin prefers K^+ over Na^+ by a factor of 100 while monensin prefers Na^+ over K^+ by a factor of 10 [10, 36].

3 Ionophore-Mediated Ion Transport in Membranes

The transport modes catalyzed by neutral and carboxylic ionophores in membranes are detailed in Fig. 3 [17].

3.1 Neutral Ionophores

The transport mode catalyzed by neutral ionophores is given on the right of Fig. 3. The ionophore, which is preferentially partitioned into the membrane, diffuses to the membrane interface (a). Here, it combines with a complexable cation initially forming a hydrophilic encounter complex (b). Replacement of the solvent molecules constituting the primary solvation sphere of the cation with the liganding moieties of the ionophore transforms the initial encounter complex into a lipophilic inclusion complex (c). The complex, which is now compatible with the apolar membrane interior, diffuses across the membrane phospholipid bilayer (d). Upon emerging at the opposite membrane interface, the complex dissociates (e). The uncomplexed ionophore is now free to diffuse back into the membrane (f). The net result is the movement of an ion M^+ together with its *charge* across the membrane.

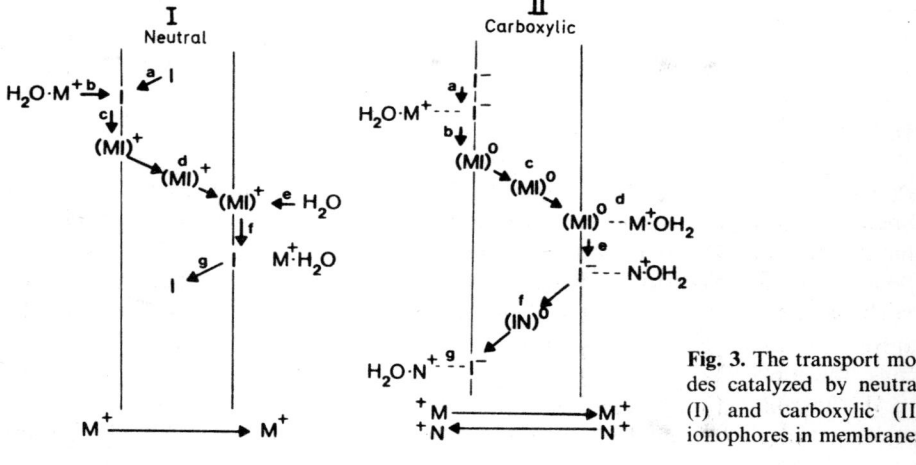

Fig. 3. The transport modes catalyzed by neutral (I) and carboxylic (II) ionophores in membranes

Since neutral ionophores essentially render the membrane differentially permeable to a positive charge (the charge carried by M^+) and since electrical potentials are actively maintained across the membranes of virtually all cells, two factors determine the thermodynamic gradient governing net transport of M^+ by neutral ionophores: (1) the magnitude of the membrane potential, i.e. ΔEm_{AB}, and (2) the concentration gradient of the permeable ion M^+ across the membrane. In terms of experimentally measurable parameters, a Nernstian relationship is established:

$$\Delta Em_{AB} = \frac{2.303\, RT}{\xi n} \log [M_A^{n+}]/[M_B^{n+}] \qquad (1)$$

where n is the charge on the transported species M. If the pre-existing transmembrane potential, Em_{AB}, exceeds the concentration term, $[M_A^{n+}]/[M_B^{n+}]$, then the ionophore induced movement of M^+ down its concentration gradient dissipates the transmembrane potential. Under these conditions the carrier is operating in an electrophoretic transport mode. If the concentration term exceeds the pre-existing potential term, the movement of M^+ down its concentration gradient will result in an increase in Em_{AB}. This is termed electrogenic transport. If gradients of more than one monovalent complexable cation exist across the membrane, the final equilibrium is described by the Goldman equation:

$$\Delta Em_{AB} = 59\text{ mV} \log \frac{[M^+]_A + P_{MN}[N^+]_A \cdots}{[M^+]_B + P_{MN}[N^+]_B \cdots} \qquad (2)$$

where P_{MN} is the relative ionophore unduced permeability ratio for M^+ to N^+. Thus, neutral ionophores not only alter transmembrane concentration gradients, but alter membrane potentials as well.

3.2 Carboxylic Ionophores

The transport cycle catalyzed by carboxylic ionophores is given on the left of Fig. 3. At a physiological pH of 7.4 the carboxylic functionality is ionized. As a result, the ionized ionophore is confined to the polar face of the membrane. When the anionic ionophore encounters a complexable cation, M^+, at the interface, complexation is initiated by formation of a solvent-separated ion pair between the hydrated cation and the anionic ionophore (a). The ionophore now garlands itself about the cation, replacing the solvents of the primary solvation sphere with its liganding moieties (b). The zwitterionic complex, its charge internally compensated, is now capable of breaking away from the polar interface towards the membrane interior. The outward orientation of the lipophilic alkyl groups on the ionophore backbone enables the complex to passively diffuse across the apolar membrane interior to the opposite membrane interface (c). As was the case with the neutral ionophores, movement of the complex into the polar region at the opposite membrane interface facilitates the release of the cation which then resolvates (d). If the anionic ionophore, now confined to membrane interface B, is able to combine with a cation, N^+, the lowering of its polarity would permit re-entry into the membrane interior (e) and return to the initial interface A (f) where N^+ would be released (g). The result is the exchange of one cation for another across the membrane with no net charge translocation. In classical terms, the carboxylic ionophores act as *exchange diffusion carriers*. Protons can substitute for the complexible cation in the reaction scheme described. Since both the protonated and zwitterionic forms of the carboxylic ionophore are neutral, the equilibrium catalyzed by these ionophores contains no electrical terms and is independent of the membrane potential:

$$[M^{n+}]_A/[M^{n+}]_B = [N^{n+}]_A/[M^{n+}]_B = [M^+]_A/[H^+]_B \ . \tag{3}$$

The condition of electrically neutral transport appears to be an essential requirement for tolerance of pharmacologically active concentrations of ionophores. Since membrane potentials play an important role in regulating excitable cells, e.g. nerve, muscles and secretory cells, the neutral ionophores which alter transmembrane electrical potentials are capable of creating metabolic chaos in cells, organs and intact animals. Consequently, they tend to be exceedingly toxic. On the other hand, carboxylic ionophores, which behave as exchange diffusion carriers by exchanging cations across membranes in electrically neutral zwitterionic complexes, induce minimal net charge translocation across the membrane. For this reason, carboxylic ionophores have considerably greater potential as therapeutic agents than their more toxic neutral counterparts. Indeed, carboxylic ionophores have been found to elicit specific cardiovascular effects [37] and have found wide application as animal feed additives [38, 39].

4 Conformational Dynamics of Ionophores

4.1 Macrocyclic Neutral Ionophores in Bulk Solvents

It was recognized quite early that the ion complexing activities of many neutral ionophores of the cyclodepsipeptide type are influenced by environmental effects on ionophore backbone conformation [40]. Optical rotatory dispersion (ORD) studies of enniatin B in a series of solvents indicated the existence of a conformational equilibrium, the position of which was solvent dependent. A single conformer, designated form N, with a strong negative Cotton effect at 230 nm predominates in nonpolar solvents [23]. Increases in solvent polarity shift the conformational equilibrium until a new conformer, form P, characterized by a weak positive Cotton effect at 240 nm and a strong negative Cotton effect at 200 nm finally predominates at high solvent polarity. The conformation of the K$^+$ complex of enniatin B is of the same general orientation as that of the non-complexed ionophore in polar solvents [23].

Extensive NMR [19,41,42], circular dichroism [19,40], and infrared studies [19,40] along with conformational energy calculations [43,44] have been used to elucidate the conformational dynamics of valinomycin which accompany cation complexation in solution and in phospholipid membranes [45]. The conformation of valinomycin is very solvent dependent and can exist in any of three forms designated A, B and C by Ovchinnikov et al. depending upon solvent polarity [20]. In form A, predominate in non-polar solvents (CCl$_4$, CHCl$_3$ and heptane-dioxane 10:1), all NH groups participate in intra molecular hydrogen bonding with the amide carbonyls. The cyclodepsipeptide chain forms a fused system of six hydrogen-bond-closed ten membered rings. Consequently, in non-polar media uncomplexed valinomycin assumes a compact conformation resembling a bracelet with all six ester carbonyls directed outward. As the solvent polarity increases, the conformational equilibrium shifts to form B which is stabilized by three internal hydrogen bonds and is more flexible than form A. Form C is devoid of all internal hydrogen bonds and exists only in high polarity solvents. This form has no fixed structures and constitutes an equilibrium mixture of many similar conformers.

4.2 Carboxylic Ionophores in Bulk Solvents

Because of the greater flexibility imparted them by the lack of head-to-tail co-valent linkage, the carboxylic ionophores respond much more strongly to environmental forces such as local polarity than do the neutral macrocyclic ionophores. Upon leaving a membrane interface during the course of a catalytic transport cycle, an ionophore does not experience an abrupt change from a polar aqueous environment to an apolar hydrocarbon-like environment. The polarity boundary is rather diffuse. In order to properly evaluate the factors affecting carboxylic ionophore mediated transport, it is necessary to determine the effects of each of the microenvironments encountered within the membrane on the conformation of the ionophore and the stability of the ionophore-ion inclusion complex.

Experiments in our laboratory have been directed toward determining the conformational options of the carboxylic ionophore lasalocid (Fig. 4) in a series of solvents chosen to model the polarity gradient encountered when traversing a biomembrane. The presence of multiple asymmetric centers on the carbon backone confers a chirality on lasalocid which can be probed by examining the circular dichroism (CD) arising from the medial C$_{12}$ ketone and the terminal aromatic C ring. By using CD to monitor the conformational changes induced by changes in solvent polarity, we are able to superimpose dynamic conformational factors onto the exchange diffusion mechanism through which carboxylic ionophores operate in biomembranes.

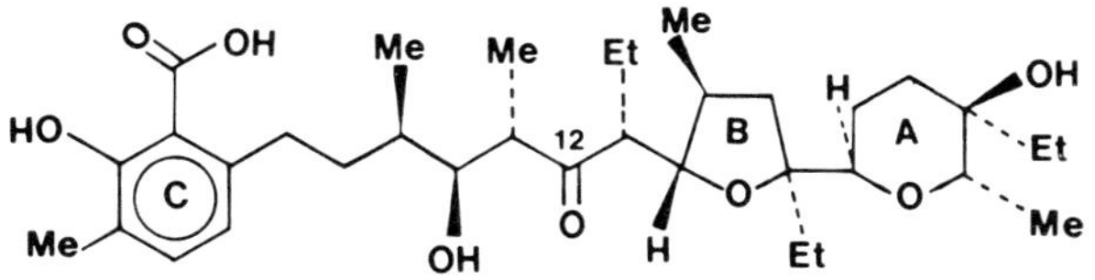

Fig. 4. Structure of the carboxylic ionophore lasalocid [46]. The CD chromophores are the medial C$_{12}$ ketone and the terminal aromatic C ring

Figure 5 presents the UV absorption and CD spectra of lasalocid. The three electronic absorption bands are characteristic of the salicylate C ring: (1) an $A_{1g} \rightarrow E_{1\mu}$ $\pi \rightarrow \pi^*$ transition at 210 nm; (2) an $A_{1g} \rightarrow B_{1\mu}$ $\pi \rightarrow \pi^*$ transition at 245 nm; (3) an $A_{1g} \rightarrow B_{2\mu}$ $\pi \rightarrow \pi^*$ transition at 317 nm (nomenclature according to Ref. [47]).

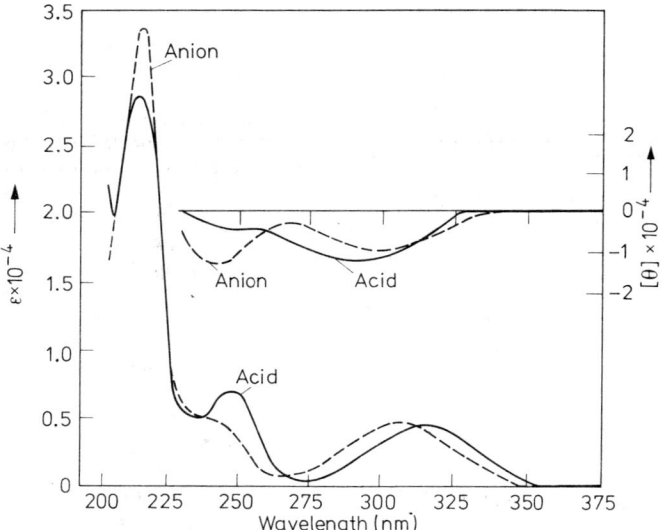

Fig. 5. The absorption and CD spectra of protonated lasalocid (———) and lasalocid anion (− − − −) in ethanol. The protonated acid form was stabilized by the addition of 0.5 equivalents of HCl. Anionic lasalocid was generated by the addition of 1.5 equivalents of tri-n-butylamine to protonated lasalocid

The 245 nm and 317 nm $\pi \to \pi^*$ transitions of the C ring give rise to conformationally sensitive CD bands at corresponding wavelengths. The 210 nm transition was not considered since it lies outside the transparency range of several of the solvent systems used to construct the polarity continuum. The position of the $A_{1g} \to B_{1\mu}$ $\pi \to \pi^*$ absorption maximum at 245 nm corresponds to the CD maximum observed at 245 nm (peak I). The CD band corresponding to the 317 nm $A_{1g} \to B_{2\mu}$ $\pi \to \pi^*$ absorption band (peak IIa) is contained along with the C_{12} ketonic $n \to \pi^*$ transition (peak IIb) in the CD composite peak at 294 nm (peak II), i.e. peak II in Fig. 5 = peak IIa plus peak IIb. Although the $n \to \pi^*$ transition of the C_{12} ketonic carbonyl is too weak to appear in the UV absorption spectrum, it nevertheless generates a strong CD band. The presence of peak IIa within peak II is easily confirmed by sodium borohydride reduction of the C_{12} ketone [48] to the corresponding alcohol which removes peak IIb from composite peak II (Fig. 6).

Since carboxylic ionophores transport ions by an electrically silent exchange diffusion mechanism, it is the anionic form of the ionophore which interacts with cations at membrane interfaces [17]. Therefore, the lasalocid species most germane for ion complexation within membranes is the free anion [49]. Deprotonation of the C_{25} carboxyl with base results in substantial changes in the absorption and CD spectra from that of the protonated form stabilized by 0.5 equivalents of HCl (cf. Fig. 5). The 245 nm absorption band observed for protonated lasalocid intensifies and shifts to 240 nm upon deprotonation while the 317 nm absorption band shifts to 310 nm with little change in intensity. In the CD spectrum, peak I shifts hypsochromically to 240 nm and intensifies upon deprotonation while peak II shifts bathochromically and diminishes in intensity.

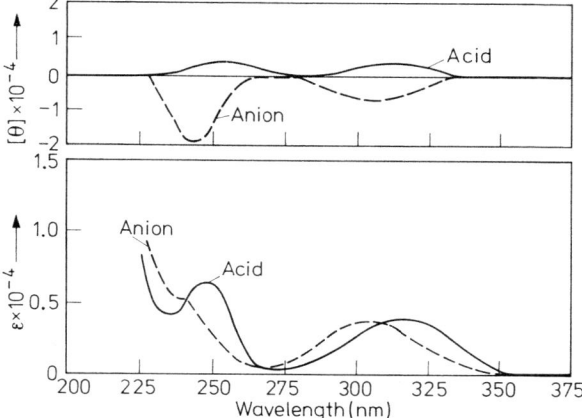

Fig. 6. The absorption and CD spectra of protonated dihydrolasalocid (———) and dihydrolasalocid anion (– – – –) in ethanol. The protonated form was stabilized by 0.5 equivalents of HCl. The anion was generated by the addition of 1.5 equivalents of tri-n-butylamine to protonated dihydrolasalocid. Peak IIa is clearly visible at 310 nm

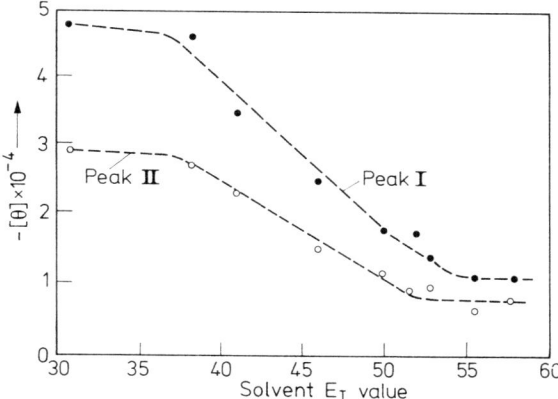

Fig. 7. $[\theta]_I$ and $[\theta]_{II}$ of lasalocid anion as a function of solvent E_T value. E_T values are effective parameters for correlating solvent properties such as the ability to donate an electron pair or to hydrogen bond with ionophore solution conformation [50, 51, 52]

The magnitudes of the CD transitions are *extremely* solvent dependent. Figure 7 shows the molar ellipticities for peaks I and II, $[\theta]_I$ and $[\theta]_{II}$, as a function of the empirical solvent polarity parameter E_T [53, 54]. $[\theta]$ decreases linearly for both peaks I and II between E_T values of 40 and 54. Above and below these values, $[\theta]$ of both peaks is relatively constant. These results are in accord with previous observations of Degani and Friedman which indicate that the conformational equilibrium of lasalocid anion varies between limiting states which depend on solvent polarity [55].

In order to ascertain the locus of the conformational changes and the structural factors which mediate them, it is necessary to resolve peak II into its IIa and IIb components. The IIb peak is a function of wavelength which can be satisfactorily approximated by a Gaussian curve of the form $[A_i \exp(-k_i(\lambda - \lambda_i)^2)]$, where A_i is the molar ellipticity $[\theta]_i$, λ_i is the wavelength of the band mean and k_i is a factor related to the standard deviation [56]. Examination of Fig. 6 shows that peak IIa for anionic dihydrolasalocid is also Gaussian and can be used to establish limiting values for the λ_{IIa} and k_{IIa} parameters of a Gaussian model of the IIa peak of lasalocid A anion. Peak II can now be resolved into the IIa ans IIb component peaks by considering it to be the superposition of two Gaussian bands:

$$II = [A_{IIa} \exp(-k_{IIa}(\lambda - \lambda_{IIa})^2)] + [A_{IIb} \exp(-k_{IIb}(\lambda - \lambda_{IIb})^2)]. \quad (4)$$

The values of A_{IIa}, A_{IIb}, k_{IIa}, k_{IIb}, λ_{IIa} and λ_{IIb} were varied independently by computer until the position, extremum and shape of the computer-generated curve matched those observed experimentally (cf. Ref. [56-58]). The relation of peak II to its component peaks IIa and IIb for one representative set of conditions is given in Fig. 8.

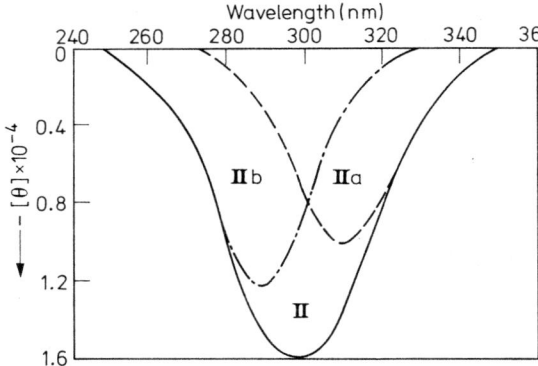

Fig. 8. Deconvoluted peak II of lasalocid anion generated in acetonitrile with 1.5 equivalents of tri-n-butylamine. Peak IIa arises from the $A_{1g} \to B_1$ $\pi \to \pi^*$ transition of the aromatic C ring and peak IIb from the $n \to \pi^*$ transition of the C_{12} ketone

$[\theta]$ values for the resolved IIa and IIb peaks are given in Table 1. Changes in $[\theta]_{IIa}$ and $[\theta]_{IIb}$ independently parallel those observed for the composite peak II. $[\theta]_{IIa}$ and $[\theta]_{IIb}$ both decrease linearly between E_T values of 40 and 54. Above and below these values $[\theta]$ is virtually constant for both peaks.

Complex formation between lasalocid anion and alkali cations results in large, solvent dependent increases in $[\theta]_I$ and $[\theta]_{II}$ (Table 2). $[\theta]$ values are consistently larger for the potassium complexes (Fig. 9). As the solvent E_T value drops the magnitudes of peaks I and II increase for both complexes.

Table 1. Resolution of Lasalocid Anion CD Peak II into Component Peaks IIa and IIb

Solvent	E_T	$-[\theta]_{II} \times 10^{-4}$	$-[\theta]_{IIa} \times 10^{-4}$	$-[\theta]_{IIb} \times 10^{-4}$
Dioxane/Water 1:1	57.9	0.81	0.42	0.66
Methanol	55.5	0.72	0.38	0.64
Dioxane/Water 8:2	52.8	1.11	0.68	0.71
Ethanol	51.9	1.08	0.65	0.85
Dioxane/Water 9:1	49.7	1.24	0.94	0.96
Acetonitrile	46.0	1.56	1.00	1.20
Dichlormethane	41.1	2.34	1.28	2.00
Chloroform	39.1	2.85	1.62	2.20
n-Hexane	30.9	2.93	1.68	2.24

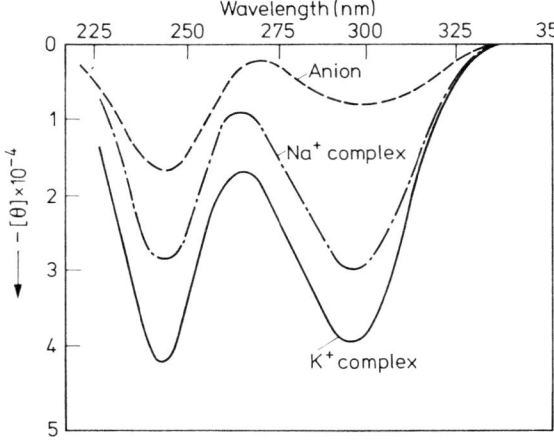

Fig. 9. The CD spectra of the Na^+ (—·—·—) and the K^+ (———) complexexes of lasalocid anion (— — — —) in ethanol. Na^+ and K^+ complexes are generated by titration of lasalocid anion to saturation with the appropriate alkali thiocyanate solution

Table 2. Effects of Alkali Cation Inclusion on the CD Spectrum of Lasalocid Anion in a Solvent Series of Decreasing Polarity

Cation	CH_3OH $-[\theta] \times 10^4$	EtOH $-[\theta] \times 10^4$	CH_3CN $-[\theta] \times 10^4$	CH_2Cl_2 $-[\theta] \times 10^4$
Na^+				
I	2.54	3.13	4.50	5.10
II	2.00	2.90	3.80	4.70
II a	0.68	0.95	1.29	1.63
II b	1.50	2.38	3.10	3.90
K^+				
I	3.40	4.80	5.20	5.36
II	3.20	4.40	4.80	5.00
II a	1.18	2.00	2.60	2.50
II b	2.74	3.80	4.20	4.50

The deconvoluted values for peaks IIa and IIb of the cation-ionophore complexes given in Table II were calculated in a manner analogous to that employed for determining the values of peaks IIa and IIb for lasalocid anion. Cation complexation results in extremely large increases in $[\theta]_{IIb}$. Changes in $[\theta]_{IIa}$ are parallel and proportional to changes in $[\theta]_I$. Although greater in absolute magnitude, changes in the CD spectrum of lasalocid anion upon cation inclusion resemble changes induced in the CD spectrum of lasalocid anion by *decreasing solvent E_T value* (compare Tables 1 and 2). The larger changes observed in $[\theta]$ upon cation complexation probably arise to some extent from contributions to the rotational strength of the chromophores due to *disymmetric placement* of the static charge of the complexed alkali ion within the ionophore complexation sphere [59]. Nuclear magnetic resonance studies indicate that the conformations of sodium lasalocid and uncomplexed lasalocid in an apolar solvent ($CHCl_3$) are virtually identical [60,61]. From the combined CD and NMR data, it appears that the conformer stabilized at low E_T values strongly resembles the conformer stabilized by alkali cation inclusion.

The structural features of lasalocid anion which mediate its conformational responses to the environment have been analyzed using computer modeling techniques [62]. Conformer structures were generated under two limiting sets of solvent-solute interactions. In a *highly polar environment* it can be assumed that solvation of all the oxygen atoms is maximal [63,64]. This condition eliminates intramolecular electrostatic interactions. In a *low polarity environment*, it can be assumed that charge neutralization via intramolecular H-bonding is paramount. The two limiting conditions primarily affect the stability of two hydrogen bonds: (1) $O_{26}^--HO_{40}$ and (2) $O_{26}^--HO_{31}$. A third intramolecular hydrogen bond, the $O_{28}H-O_{27}$ salicylate hydrogen bond, is extremely stable and polarity independent. The occurrence and stability of these hydrogen bonds in solution has been confirmed with NMR by monitoring the exchangeable hydroxyl protons as a function of solvent polarity [60].

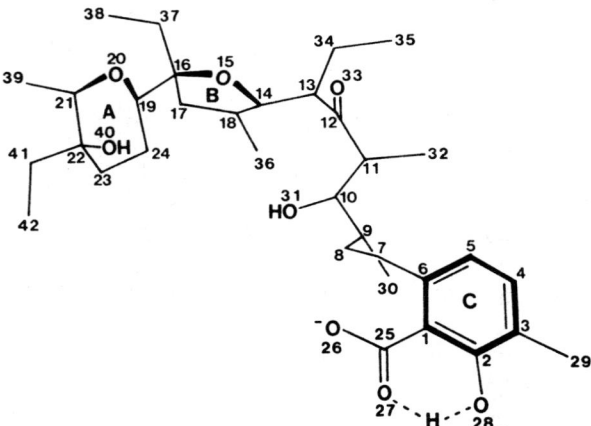

Fig. 10. The acyclic quasilinear conformer of lasalocid anion which predominates under high polarity conditions. This conformer has the lowest intrinsic molecular strain energy as calculated using a QCFF energy minimization program [65, 66]. Solvent effects on the conformation were modeled using an empirical specific site approach [64]

Under high polarity conditions, energy minimization of lasalocid anion by computer modeling yielded the conformation of lowest intrinsic molecular strain energy, the acyclic quasilinear conformer (Fig. 10). The tetrahydropyran A ring is in a chair conformation. The C_{22} ethyl group and the $C_{16}-C_{19}$ bond joining ring A with ring B are equitorial, while the C_{22} hydroxyl group ($O_{40}H$) and the C_{21} methyl group are trans diaxial. The tetrahydrofuran B ring is in an envelope conformation with C_{18} lying below the plane of the other four atoms (C_{14}, O_{15}, C_{16} and C_{17}). The aromatic C ring is essentially planar. The C_{25} carboxylate, however, is twisted out of the plane of the ring very slightly (3°) to minimize unfavorable steric interactions with the ionophore backbone at C_6. The carbon-carbon backbone bonds lying between rings B and C, C_6 through C_{14}, are all in their lowest energy, staggered conformation with torsion angles near 180°.

Formation of the cyclic conformer (Fig. 11) under apolar conditions proceeds without significant conformational changes in rings A or B. The aromatic C ring retains its planarity. The carboxylate, however, twists 26° out of the plane of the C

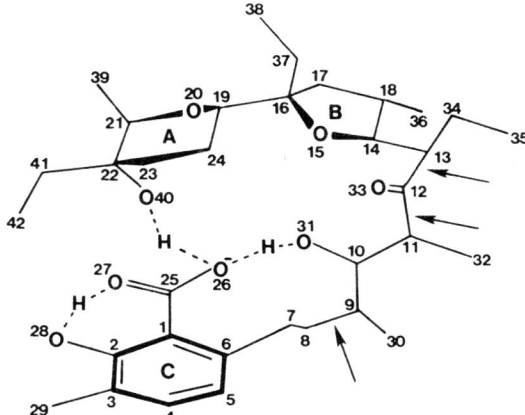

Fig. 11. The cyclic conformer of lasalocid anion which predominates under low polarity conditions and in alkali cation complexes. Rearrangement from the acyclic to cyclic conformer proceeds principally by rotation about the C_8-C_9, $C_{11}-C_{12}$ and $C_{12}-C_{13}$ hinge bonds (indicated by arrows). Dashed lines (— — —) indicate intramolecular hydrogen bonds. The hydrogen bonds help to stabilize the cyclic conformer

ring during formation of the $O_{40}H\text{-}O_{26}^-$ hydrogen bond due to the steric bulk of the carbon backbone at the C_6 ortho position. The major conformational changes during cyclization occur by rotation around the $C_8\text{-}C_9$, $C_{11}\text{-}C_{12}$ and $C_{12}\text{-}C_{13}$ carbon-carbon bonds. The $C_7\text{-}C_8\text{-}C_9\text{-}C_{10}$ torsion angle rotates from *anti* to *gauche* which moves the C_{25} carbocylate into the vicinity of the $O_{31}H$ hydroxyl and O_{33} ketonic liganding moieties. The carboxylate is now in a position to form the $O_{26}^-\text{-}HO_{31}$ intramolecular hydrogen bond. Cyclization is completed by reduction of the $O_{33}\text{-}C_{12}\text{-}C_{11}\text{-}C_{10}$ torsional angle from 74° to 32° which focuses the AB ring system into the central liganding cavity and places the $O_{40}H$ hydroxyl in a position to form the second structure stabilizing hydrogen bond, $O_{40}H\text{-}O_{26}^-$. Van der Waals strain arising from 1,3 steric interactions between the C_{32} methyl group and the C_{34}, C_{35} ethyl group is minimized by a decrease in the $O_{33}\text{-}C_{12}\text{-}C_{13}\text{-}C_{14}$ torsional angle from 60° to 29°. These three bonds play the role of the *hinge* bonds previously proposed for A23187 [67,68]. Resolvable NMR proton-proton coupling constants for lasalocid do not change upon cyclization confirming that the conformational changes accompanying cyclization proceed distal to the $C_9\text{-}C_{10}$, $C_{10}\text{-}C_{11}$, $C_{13}\text{-}C_{14}$ and $C_{14}\text{-}C_{18}$ bonds [60,61].

The conformational changes associated with cyclization are energetically unfavorable; the computer model indicates that $\Delta G_{1\text{Conform}}$ is ca. 12 kcal/mole. The unfavorable cyclization energy is generated principally by torsional strain introduced by twisting about the hinge bonds, and non-bonded, through-space interactions between the C_{32} methyl and C_{34}, C_{35} ethyl group. As solvent polarity drops, however, the stabilization of the carboxylate gained by formation of the intramolecular hydrogen bonds, and the growing propensity for hydrophobic over hydrophilic interactions, overcome the conformational energy barrier and the molecule cyclizes.

Correlation of solvent and cation inclusion induced changes in the CD spectra with the computer generated molecular models supports the cyclization mechanism [51,58]. The rotation of the C_{25} carboxylate accompanying cyclization destroys the plane and center of symmetry of the aromatic chromophore rendering it *dissymmetric*. Induction of dissymmetry into the aromatic chromophore establishes a diastereomeric relationship with the existing stereochemistry of the ionophore backbone. Since the backbone is capable of bending such that approach of the $O_{40}H$ to form the head-to-tail hydrogen bond with the C_{25} carboxylate is more favored from *above* the plane of the C ring (+ rotation of the carboxylate) than from *below* (− rotation of the carboxylate), one of the two possible diastereomers predominates. The dissymmetric contributions to the aromatic $\pi \to \pi^*$ transitions arising from the preponderance of the single diastereomer should be quite large and greatly outweigh those induced by asymmetrically disposed substituents [69,70]. Consequently, the magnitude of the $\pi \to \pi^*$ bands will be dominated by the stereochemistry in the region of the carboxylate [59,69,70]. The *large* changes in $[\theta]_I$ and $[\theta]_{IIa}$ caused by a solvent polarity drop (cf. $E_T = 30.9$ (hexane) and $E_T = 55.5$ (MeOH) in Table 1) or by alkali cation inclusion (see Table 2) are consistent with the induction of dissymmetry into the aromatic chromophore.

Further confirmation results from the resolution of peak IIb from composite peak II which allows the direct observation of solvent polarity and cation inclusion effects on the sign and magnitude of the CD signal arising exclusively from the C_{12} $n \to \pi^*$ transition [51]. Two of the three principal rearrangements associated with cyclization occur adjacent to the C_{12} ketone, which consequently provides a sensitive CD probe for monitoring cyclization. Peak IIb becomes increasingly negative as the solvent polarity drops (see Table 1) or upon inclusion of an alkali ion (see Table 2). Application of the *octant* rule to the computer generated models provides a semi-quantitative prediction of the effect cyclization should have on $[\theta]_{IIb}$ [71,72]. Cyclization moves the AB ring system from the front lower left and the front upper left octants fully into the front upper left octant and thus should result in an overall increase in the magnitude of the negative IIb transition, as is indeed observed for the solvent polarity decrease and for cation inclusion.

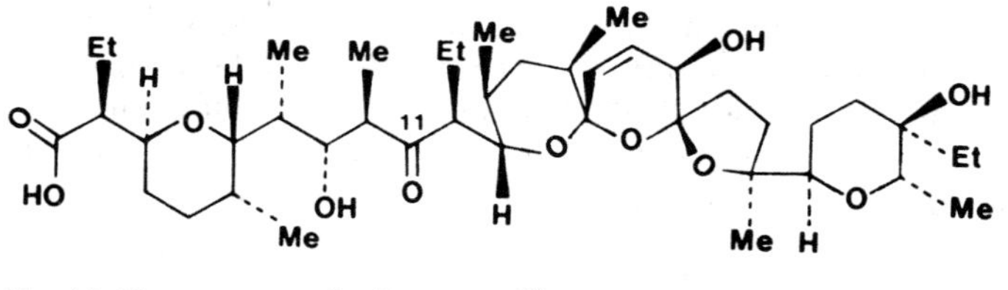

Fig. 12. The structure of salinomycin [73]

Analogous CD studies of the effects of solvation and cation inclusion on the conformation of salinomycin (Fig. 12) reveal a solvent mediated equilibrium similar to that observed for lasalocid [50,51]. Salinomycin anion tends toward one of two metastable conformational states depending upon solvent polarity. In solvents of high polarity, $E_T > 55$, the $[\theta]$ value of the $n \to \pi^*$ CD transition is constant and low. In a narrow range of intermediate solvent polarity corresponding to E_T values between 55 and 48, there is an abrupt shift in the conformational equilibrium. Below an E_T of 48, $[\theta]$ is constant and of considerably higher magnitude. Preliminary analysis indicates that both conformers of salinomycin are single turn helices dictated by the constraints of the tricyclic spirane system and that lowering polarity shifts the conformation from a wide pitch helix to a tight pitch helix. Although the solvent polarity dependence of the conformation of salinomycin anion basically parallels that of lasalocid anion, the polarity range over which the conformational equilibrium shifts is different. These results suggest that structural features unique to each carboxylic ionophore are ultimately responsible for the response of the ionophore to environmental factors, e.g. solvent polarity or locus within a membrane.

4.3 Carboxylic Ionophore Conformational Dynamics During Transport

Data derived from the solution studies of lasalocid permits dynamic conformational factors to be superimposed on the presently accepted mechanism of lasalocid-mediated cation transport which presumably are applicable to carboxylic ionophore-mediated cation transport in general. The transport cycle begins with the lasalocid A anion in its extended *acyclic* conformation confined to the membrane interface where it is stabilized by the polar environment in the lipid polar head group region (the polarity of the lipid polar head group region appears to be similar to that of methanol, i.e. $E_T \geq 55.5$) [74]. Under these conditions, the liganding heteroatoms are strung out and consequently cannot present a concerted inducible dipole system capable of strong interaction with solution cations. The strength of a cation-dipole interaction between a given backbone liganding oxygen, e.g. O_{31}, O_{33}, O_{15}, O_{20}, O_{40} for lasalocid A, and a solution cation is inversely proportional to the *cube* of their separation distance while the strength of a cation-anion interaction between the C_{25} carboxylate and a solution cation is inversely proportional to the *square* of their separation distance [75]. Thus, ion pairing rather than ion-induced dipole interaction appears to be the most likely mechanism for initial interactions between the acyclic ionophore at the membrane interface and solution cations. Ion pairing of this type represents the limit of interaction attainable between the acyclic ionophore conformer and hindered cations such as the alkylammonium ions [58]. If equivalent ion pairs are formed with smaller alkali cations, localization of the cation on the C_{25} carboxylate makes ion-induced dipole interactions more favorable with neighboring liganding oxygens and the resulting forces reorient the ionophore backbone about the *hinge* bonds into the *cyclic conformer*. As molecular reorientation proceeds, the heteroatoms move closer to the optimal ligand-cation bond distance permitted by structural constraints and form a liganding field capable of stabilizing the cation relative to the bulk solvent. The cyclic conformer with the polar liganding groups focused into the cation binding cavity and the lipophilic alkyl groups shielding the exterior is the form most compatible with the apolar membrane interior. Since this conformation has a reduced capability to interact with the polar environment at the membrane surface, it readily leaves the interface and enters the lipophilic membrane interior where it is further stabilized. Upon diffusion to the opposite membrane face, the complex is again subjected to interaction with a polar environment. Since electrostatic stabilization no longer supersedes the unfavorable ΔG of cyclization, the complex re-equilibrates with the polar environment releasing the cation and depositing the acyclic anion at the interface for the next phase of the transport cycle.

5 Specific Membrane-Ionophore Interactions

5.1 Membrane Composition and Ionophore Activity

Although a great deal has been learned about the mechanism of ionophore mediated transport by studies conducted in free solution, evidence suggests the solution environment model of ionophore complexation to be an oversimplification and that specific, activity-altering interactions exist between ionophores and the polar head groups of lipid membranes. Direct interaction of the carboxylic ionophore lasalocid with membranes has been detected as an enhancement of the intrinsic fluorescence arising from the salicylate residue on the terminus of the ionophore [74]. Using the fluorescence emission enhancement which accompanies movement of the salicylate chromophore into the less polar membrane environment, Haynes and Pressman calculated the binding constant of lasalocid A to monolayer dimyristoyl phosphatidylcholine (DMPC) vesicles (ca. 500 Å diameter spheres of di-n-butylether covered with a monolayer of DMPC) to be 1×10^{-4} M^{-1} at a pH of 7.3 [74].

Fluorescent lifetime experiments have shown lasalocid to exist in two different environments when bound to vesicle membranes, possibly reflecting two different modes of ionophore-phospholipid interaction [49]. The fluorescent lifetimes within the two different environments differ by a factor of approximately two. The longer lived species is calculated to compose between 11 and 20% of the total bound ionophore. Several explanations for the existence of two bound species have been put forward [49]: (1) an inward versus outward orientation of the fluorescent salicylate group; (2) perpendicular versus parallel orientation of the ionophore with respect to the plane of the membrane; (3) deep versus shallow insertion of the salicylate group with respect to the glycerol bridge region of the membrane; (4) modes of binding resulting in lesser or greater environmental or ionophore mobility.

The total lasalocid-membrane binding reaction is a very sensitive function of the membrane polar head group composition [74]. Incorporation of dimyristoyl phosphatidic acid (DMPA) or dimyristoyl phosphatidyl ethanolamine (DMPE) into pure DMPC vesicles reduces the total lasalocid fluorescence relative to that observed in pure DMPC vesicles. Fluorescence lifetime experiments indicate the decrease in fluorescence is not due to reduction of the quantum yield of the bound lasalocid but rather to a reduction in the degree of binding. The sensitivity of total ionophore binding to polar head group composition is similar to that observed for the fluorescent spin probe 1-anilino-8-naphthalenesulfonate (ANS^-) [76, 77] and indicates that anionic lasalocid, like ANS^-, is excluded from the hydrocarbon interior of the membrane and confined to the polar head group region at the membrane interface.

Membrane composition also has profound effects on the cation complexation activity of lasalocid within vesicle membranes [49]. In pure DMPC monolayer vesicles, lasalocid displays an apparent complexation constant for Ca^{2+}, $K_{app}(Ca^{2+})$ of 1423 M^{-1} and an apparent complexation constant for K^+, $K_{app}(K^+)$ of 52.5 M^{-1}. When the lipid composition of the vesicle membrane is changed to 50% DMPC/50% DMPA the $K_{app}(Ca^{2+})$ plumets to 40.1 M^{-1} while the $K_{app}(K^+)$ drops less drastically to 33.0 M^{-1}. Pure DMPA monolayer vesicles display a $K_{app}(Ca^{2+})$ of ca. 3000 M^{-1} and a $K_{app}(K^+)$ of 15.7 M^{-1}.

The intrinsic fluorescence of A23187 is a sensitive probe of the polarity of the ionophore environment and divalent cation binding state [78,79]. The mechanism of divalent cation transport by A23187 across phospholipid bilayers has been investigated using the rate of fluorescence changes observed in stopped-flow kinetic experiments [80]. An apparent rate constant (k_{app}) was determined for Ca^{2+} : ionophore transport across vesicle bilayer membranes and was found to vary with the lipid composition of the membrane. Maximal values of k_{app} were observed for vesicles prepared from pure DMPC ($k_{app} = 0.98$ sec^{-1}). Inclusion of 33% DMPE ($k_{app} = 0.217$ sec^{-1}), 31% DMPA ($k_{app} = 0.057$ sec^{-1}) or 32% dipalmitoyl PC (DPPC) ($k_{app} = 0.019$ sec^{-1}) resulted in lower values of k_{app}.

The changes in k_{app} induced by changes in lipid composition can be attributed to changes in ionophore-polar head group interactions and/or changes in membrane fluidity. Temperature dependence studies in DMPC membranes indicate however, that k_{app} is virtually independent of membrane fluidity [80]. A fit of k_{app} for Ca^{2+} : A23187 transport across DMPC membranes to temperature is linear over a temperature range that includes the lipid phase transition temperature (ca. 23 °C for DMPC [81]) indicating that dramatic changes in membrane fluidity *do not affect* transport. Kolber and Haynes [80] offer two possible explanations for the lack of fluidity dependence: (1) the ionophore fits into both lipid phases (crystalline and gel) equally well or (2) the ionophore disrupts the environment to such an extent that it is insensitive to gross organizational changes in the hydrophobic region of the membrane.

The lack of sensitivity of A23187 induced Ca^{2+} translocation to DMPC membrane fluidity suggests that differences in transport rates encountered upon inclusion of DMPE and DMPA into DMPC vesicles is the result of ionophore-lipid head group interactions. Incorporation of 33% DMPE into pure DMPC vesicles lowers the k_{app} for Ca^{2+} translocation by 0.763 sec^{-1}. PE produces a smaller area per lipid head within a membrane bilayer resulting in a more tightly packed membrane. Tighter packing may exclude A23187 from the lipid polar head region thereby decreasing the amount of ionophore available for Ca^{2+} inclusion and transport. It has been speculated that formation of a terniary A23187$^-$-Ca^{2+}-Pa$^-$ complex in DMPC/31% DMPA vesicles could serve to reduce the amount of A23187 available within the membrane and thereby account for the reduced k_{app} observed upon incorporation of DMPA [80]. Incorporation of DPPC (32%) into DMPC vesicles greatly alters the architecture of the membrane interface. The polar head groups of DMPC and DPPC do not lie in the same plane due to the difference in acyl chain length (myristoyl, C = 14; palmitoyl, C = 16 [82]). The decrease in k_{app} upon addition of DPPA to pure DMPC vesicles may arise from more extensive membrane binding of A23187 due to the greater surface area created at the interface by the biplanarity.

5.2 Specific Carboxylic Ionophore-Phospholipid Interactions

As of yet the molecular details of the interaction mechanism between carboxylic ionophores and the polar phospholipids have not been established. Work in our laboratory on the interaction of lasalocid A and alkylammonium ions too bulky to form true inclusion complexes suggests that ion pair formation between the terminal carboxylate of the ionophore and the alkylammonium groups of PC and PE polar head groups may be a viable mechanism for the observed binding of lasalocid to PC and mixed PC/PE vesicle membranes [58]. A CD pilot study was conducted to determine if DMPC would ion pair to lasalocid A in free solution. Treatment of lasalocid A anions (generated in ethanol with tetra-n-butylammonium hydroxide) with DMPC leads to dramatic increases in the $\pi \to \pi^*$ CD transitions of the aromatic C ring without concomitant changes in the $n \to \pi^*$ CD transitions of the medial C_{12} ketone. The fact that the C_{12} ketonic transition is unaffected indicates that the ionophore backbone is not involved in the interaction. The confinement of the spectral changes to the $\pi \to \pi^*$ aromatic transitions indicates the phospholipid to be interacting solely with the C ring, presumably by an ion exchange equilibrium (Fig. 13). Changing the fatty acid chains from myristoyl to palmitoyl has no effect on the CD spectrum.

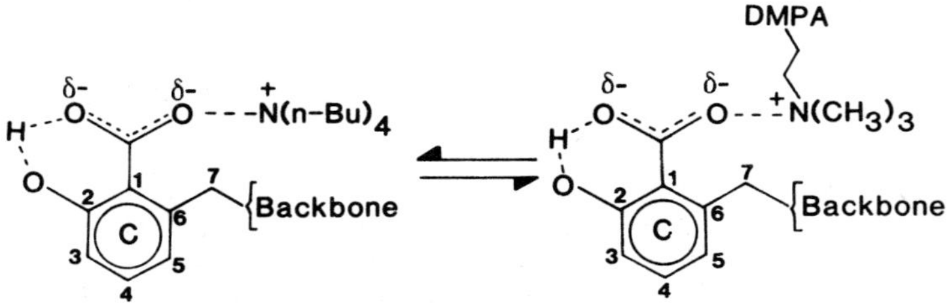

Fig. 13. Lasalocid anions generated with $(n-Bu)_4$ NOH in 80% dioxane/water are initially ion paired to $(n-Bu)_4$ N^+. Addition of another ion, DMPC, sets up an ion exchange equilibrium which is detected by changes in peaks I and IIa of the CD spectrum

6 Biological Properties of Ionophores

Both neutral and carboxylic ionophores have been extensively employed as tools for *in vitro* studies of biological systems (for reviews see refs. 17 and 83). However, because their electroneutral exchange-diffusion mode of transport does not perturb membrane potential, only the carboxylic subclass of ionophores is sufficiently tolerated by intact animals to produce well defined pharmacological responses.

6.1 Cardiovascular Properties of Carboxylic Ionophores

The ability of lasalocid and A23187 to transport Ca^{2+} was confirmed by their ability to release Ca^{2+} down its concentration gradient from loaded vesicles prepared from muscle sarcoplasmic reticulum [78]. These observations suggested that the ionophores might provide novel tools for rationally altering the ion gradients which control biological systems. Administration of lasalocid to artificially perfused, isolated guinea pig and rabbit heart preparations did indeed increase contractile force and spontaneous beat frequency [84, 85, 86]. Studies on the cardiovascular effects of lasalocid in normal anesthetized dogs showed lasalocid to produce an even more dramatic augmentation of myocardial contractility with up to a tripling of the contractility

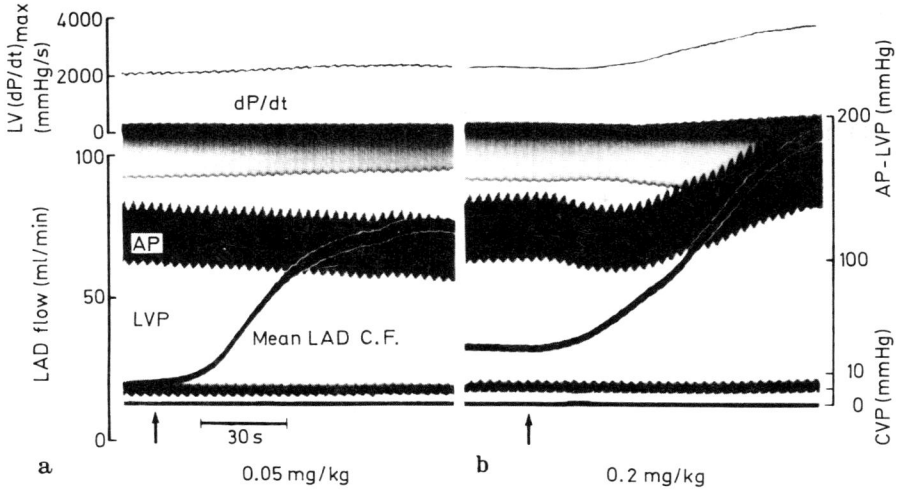

Fig. 14. Cardiovascular response of a typical anesthetized dog to monensin. A low dose (0.05 mg/kg) was first introduced i.v. (dissolved in ethanol) and after an interval of an hour to permit the animal to return to basal conditions, a higher dose (0.2 mg/kg) was administered. The lowest tracing (mean LAD C.F.) is the time averaged flow measured by a magnetic flow probe encircling the left anterior descending coronary artery. The AP trace gives the diastolic-systolic pressure range recorded from a catheter in the aorta. LV dP/dt$_{max}$, the index of cardiac contractility, was obtained from a manometer-tipped catheter inserted in the left ventricle. The measured pressure was converted to its derivative to record dP/dt directly

index, LV dP/dt, after intravenous doses of up to 2.0 mg/Kg [87]. Initial studies on the cardiovascular properties of A23187 by Schaffer et al. [88] utilizing perfused heart preparations and Schwartz et al. [86] utilizing isolated atria and perfused ventricles failed to demonstrate any significant effects on the contractile parameters of atrial or ventricular myocardium. However, a concentration dependent, positive inotropic effect has been reported in isolated guinea pig left atria [89]. Administration of 1×10^{-6} M to 3×10^{-5} M A23187 produced increases in both the force of contraction and the rate of tension development in the atria. A23187 gives only sporatic inotropic response in the intact dog [90].

The strictly monovalent ionophores, e.g. monensin, produce cardiovascular effects in animals similar to those produced by the divalent ionophores lasalocid and A23187 [37]. Figure 14 illustrates the two distinct, dose separable, primary cardiovascular effects produced by monensin. At low concentrations, 50 µg/Kg or less, monensin produces a direct dilatation, i.e. relaxation of the smooth muscle of the coronary arteries, manifested by a multifold increase in coronary blood flow. At this dosage, no other effects are apparent for a considerable period; between 2–25 µg/Kg the response is totally coronary specific. If the dose is increased to 0.2 mg/Kg, an inotropic response (an increase in cardiac contractility) can be monitored as the maximum rate of rise of LV dP/dt. Other parameters parallel the inotropic effect. Following an initial drop caused by dilatation of the systemic arteries, mean blood pressure rises as does pulse pressure, the interval between lowest (diastolic) and highest (systolic) transient pressure. The rate of blood pumped by the heart (cardiac output) also rises.

6.2 Mechanism of Action of Carboxylic Ionophores

The recognition that the strictly monovalent ionophores produce cardiovascular effects in animals similar to those produced by lasalocid indicates that the *a priori* rationale implicating direct Ca^{2+} transport by ionophores to be naive [37, 84, 85, 86]. This conclusion is reinforced by the observation that A23187, the most specific Ca^{2+} transporter, produces relatively week cardiovascular effects in intact animals [90]. It appears that the common denominator for cardiovascular responses is ionophore mediated Na^+ and K^+ transport rather than direct transport of Ca^{2+} and/or catecholamines!

Table 3 compares the *in vitro* ion carrying capacity of a series of carboxylic ionophores with their inotropic potency [91, 92]. Appreciable rates of Ca^{2+} or catecholamine (norepinephrine) transport are observed only for lasalocid, the ionophore of the group with the poorest inotropic potency. Extremely wide ranges of Ca^{2+} and norepinephrine transport capacity are seen with no correlation to inotropic response. The correlation between inotropic potency and *in vitro* Na^+ transport is less negative [37]. When the activities of the ionophores are compared on the basis of their ability to release K^+ from erythrocytes, chiefly in exchange for Na^+, the correlation with inotropic potency is even better.

Table 3. Comparison of Inotropic Potency of Ionophores with *in vitro* Transport Properties

Ionophore	Inotropic Potency	In Vitro Ca^{2+} Transport	In Vitro Norepinephrine Transport	In Vitro Na^+ Transport	Erythrocyte K^+ Release
Lasalocid	(1.0)	(1.0)	(1.0)	(1.0)	(1.0)
Lysocellin	1.5	—	—	—	4.1
Septamycin	2.0	—	—	—	—
Nigericin	2.8	.000009	0.001	1.4	16.4
Dianemycin	4.3	.00015	.1	27	18.0
Monensin	6.1	.000009	.003	31	7.2
X-206	7.7	.000025	.002	2	10.9
Salinomycin	12.1	—	—	—	10.0
A-204	13.1	.000025	.01	20	41
A-23187	±	.37	low	.002	—

Inotropic potenticies are used compared as the inverse of the ionophore dose required to double LV max dP/dt. Ca^{2+}, norepinephrine and Na^+ transport rates were obtained in the vertically stacked three phase system described in ref. [37]. Erythrocyte K^+ release potency was measured as the inverse of the concentration required to release 10 mM K^+ from washed human erythrocytes suspended in mock plasma containing 5 mM KCl, 145 mM NaCl and 10 mM TRIS chloride, pH 7.4.

There is an extensive literature implicating the indirect augmentation of intracellular Ca^{2+} activity by an increase in the concentration of intracellular Na^+. The cardiac glycosides, for example, are believed to increase intracellular Na^+ by inhibiting the Na^+ pump which, in some fashion which is not completely clear, increases the magnitude of the wave of Ca^{2+} activity imposed on the contractility receptor, the troponin C component of the actomyosin complex. This presumably involves come competition between Na^+ and Ca^{2+} for entry into sequestered intracellular Ca^{2+} pools such as relatively unspecific membrane binding sites.

Secretion is another Ca^{2+} mediated cellular activation process sharing many points in common with muscular contractility [93]. We have verified that a specific monovalent ionophore, salinomycin, does in fact cause a multifold augmentation of plasma catecholamines in animals [94]. In addition, monensin causes a release of catecholamines from cultured adrenal medullary chromaffin [95,96,97]. These observations verify earlier implications of the involvement of catecholamines in the ionophore induced cardiovascular effects of both lasalocid and monovalent selective carboxylic ionophores which were based on partial inhibition by the β-adrenergic catecholamine inhibitors such as propranolol [37,88]. Presumably, the residual, propranolol-resistant cardiovascular effects can be attributed to mobilization of intracellular Ca^{2+} activity by increased intracellular Na^+. Thus, ironically, the modulation of Ca^{2+} and catecholamines originally anticipated for lasalocid does in fact obtain for *all monovalent caboxylic* ionophores, albeit by a different, less direct mechanism.

6.3 Transport of Na^+, K^+ and H^+ Across RBC Membranes

Because of the transcellular ion gradients actively maintained across biomembranes under physiological conditions (intracellular K^+ high, Na^+ low; extracellular K^+ low, Na^+ high), the monovalent carboxylic ionophores principally promote an exchange of internal K^+ for external Na^+. However, the cation exchange is accompanied by transient changes in the environmental pH, the direction and magnitude of which depend on the relative K^+ and Na^+ affinities of the ionophore. The involvement of protons in the Na^+ for K^+ exchange catalyzed by carboxylic ionophores across erythrocyte [95,99,100] and resealed ghost membranes [101] has been known for some time.

Previous attempts to dynamically monitor the complete Na^+, H^+, K^+ exchange process mediated by carboxylic ionophores have been hampered by the lack of a direct means of monitoring Na^+ fluxes. Due to the recent availability of a stable, highly selective Na^+ glass electrode, we have re-examined ionophore mediated ion translocation across the membranes of human red blood cells (RBCs). Washed RBCs are suspended in mock serum electrolyte consisting of 140 mM NaCl, 5 mM KCl and 25 mM HEPES buffer at pH 7.4. Following the addition of ionophore, ion fluxes are monitored by Na^+, H^+ and K^+ electrodes.

Figure 15 shows the effect of addition of the K^+ selective ionophore nigericin (2.5×10^{-5} M) on a suspension of RBCs. Following addition of the ionophore, there is an immediate egress of K^+ which is countered by a stoichiometric influx of H^+ and Na^+. As a consequence, the K^+ transmembrane gradient is quickly dissipated while the Na^+ activity gradient lingers. In order to facilitate complete dissipation of the Na^+ gradient, the direction of ionophore mediated H^+ flux *reverses* to counter continued ingress of Na^+. Thus, there is a *bidirectional* proton flux associated with nigericin mediated K^+ for Na^+ exchange which is a difference function of K^+ egress and Na^+ ingress.

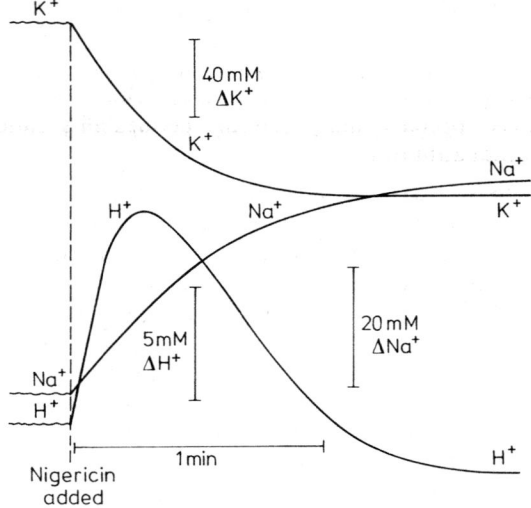

Fig. 15. Exchange of internal potassium ions for external sodium ions and protons catalyzed by nigericin, 10 µM/ml, applied to a suspension of human erythrocytes (10% packed cell volume). The changes in the concentration of sodium, potassium and hydrogen ions were monitored by ion specific electrodes. Movement of ions from the cells into solution is shown by a downward deflection

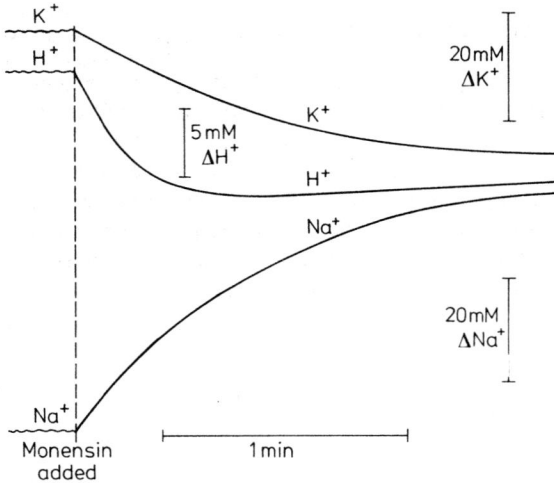

Fig. 16. Monensin, 10 µM/ml, mediated exchange of internal potassium ions and protons for external sodium ions in a suspension of human erythrocytes (10% packed cell volume). Movement of ions from the cells into solution is shown by a downward deflection

Figure 16 shows the effect of the Na^+ selective carboxylic ionophore monensin ($Na^+:K^+$ selectivity of $\approx 10:1$) on an RBC suspension. Note that protons move out of the cell with K^+ to counter an ingress of Na^+. The movement of H^+ out of the cell along with K^+ is a reflection of the propensity of monensin to select for Na^+ over K^+.

The superposition of the equilibrium affinity ratios of monensin and nigericin over the membrane permselectivity which they induce is in accord with their operating within the "equilibrium domain" [10, 102, 103]. Since the complexation-decomplexation reactions between carboxylic ionophores and cations occur rapidly enough to be considered at equilibrium relative to the rate of movement of the ion-carrier complexes across the membrane interior, the ionophore induced permeabilities simply reflect the product of the equilibrium ion selectivity intrinsic to the ionophore and the mobility ratios of the ionophore-cation inclusion complexes. Because the monovalent cation inclusion complexes of most carboxylic ionophores are virtually isomorphic [104], the mobility ratios of the complexes approach unity and the permeability ratio consequently closely reflects equilibrium affinities.

Table 4. Ionophore Mediated Ion Exchange in Human Erythrocytes[a]

Ionophore	Initial Rates (µM/min)			$K^+:Na^+$ Selectivity
	dNa^+/dt	dK^+/dt	dH^+/dt	
X-206	50	−120	70	2.40
Nigericin	66	−134	68	2.03
Lasalocid	72	−104	32	1.44
Salinomycin	20	−28	8	1.40
A-204	44	−52	8	1.18
Lysocellin	154	−140	−14	0.91
Dianemycin	174	−152	−22	0.87
Monensin	54	−20	−34	0.37

[a] Rate measurements were made on washed human erythrocytes suspended in mock serum electrolyte consisting of 140 mM Na^+, 5 mM K^+ and 20 mM HEPES buffer at pH = 7.4. Egress (−) and ingress (+) rates for K^+ and H^+ were measured directly using ion specific electrodes. The rate of Na^+ movement was calculated as a function of K^+ and H^+ under the condition of electroneutral transport [100].

Table 4 gives the initial rates for ionophore mediated Na^+, K^+ and H^+ exchange across erythrocyte membranes for a series of carboxylic ionophores [100]. The $K^+:Na^+$ selectivity intrinsic to the ionophore in the erythrocyte membrane is calculated as the ratio of $dK^+/dt:dNa^+/dt$. For all ionophores which show Na^+ selectivity, K^+ egress is accompanied by an initial H^+ egress while for ionophores showing K^+ selectivity, K^+ egress is countered by an initial H^+ ingress. The role of the proton movement in the pharmacological mechanism of action of the carboxylic ionophores is presently unknown.

In addition to the naturally occurring carboxylic ionophores listed in Table 4, we have attempted to induce ion transport with two synthetic carboxylic ionophores. The first, an α-carboxy, ω-hydroxy oligoethylene glycol derivative [105] failed to initiate detectable transport even at concentrations a hundred fold greater than the concentration of natural ionophore required to produce maximal transport. This same observation held true for a 1,1'-dinaphthyl[18]-crown-6 derivative substituted in the 3,3' position with oligoethylene glycol sidearms which terminated with carboxylic acid moieties [29].

7 Conclusion and Future Prospects

In this chapter, the basic, dynamic molecular properties underlying the transport activity of ionophores are discussed with emphasis on the effects of membrane microenvironment on the transport mechanism. Preliminary results indicate that membranes of differing phospholipid composition show differential affinities for a given ionophore as well as differential cation transport efficiencies for each equivalent of ionophore bound. The next step in understanding ionophore-mediated membrane transport is to determine the exact nature of the interactions between ionophores and membranes and how these interactions modify the dynamics of the transport process. This information will provide guidelines for the attenuation of membrane permeability by the synthesis of appropriately designed ionophores whose membrane binding characteristics and cation selectivities are targeted for specific membranes of a specific phospholipid composition. Ultimately, this will enable us to select ionophores with optimal properties for promoting a desired biological or pharmacological effect within complex membrane systems such as intact animals.

8 References

1. McMurry, W., Begg, R. W.: Arch. Biochem. Biophys. *84*, 546 (1959)
2. Moore, C., Pressman, B. C.: Biochem. Biophys. Res. Commun. *15*, 562 (1964)
3. Pressman, B. C.: Specific Inhibitors of Energy Transfer, in: Energy Linked Functions of Mitochondria (B. Chance (Ed.)), Academic Press, New York, 1963, p. 181
4. Racker, E.: Function and Structure of the Inner Membrane of Mitochondria and Chloroplasts. Amer. Chem. Soc. Monogr. *165*, 127 (1970)
5. Pressman, B. C.: Proc. Natl. Acad. Sci. U.S. *53*, 1076 (1965)
6. Pressman, B. C., Harris, E. J., in: Seventh Intern. Congr. Biochem., Tokyo, August, 1967, 5:900
7. Graven, S. N., Estrado-O, S., Lardy, H. A.: Proc. Nat. Acad. Sci. U.S. *56*, 654 (1966)
8. Mueller, P., Rudin, D. O.: Biochem. Biophys. Res. Commun. *26*, 398 (1967)
9. Lev, A. A. and Bujinsky, E. P.: Tsitologiya *9*, 102 (1967)
10. Pressman, B. C. et al.: Proc. Natl. Acad. Sci. *58*, 1949 (1967)
11. Ovchinnikov, Yu. A., Ivanov, V. T., Shkrob, A. M.: Membrane Active Complexes, B. B. A. Library, Vol. 12, Elsevier, New York 1974
12. Pauling, L.: The Nature of the Chemical Bond, New York, Cornell University Press, 1960
13. Morf, W. E., Simon, W.: Helv. Chim. Acta *54*, 2683 (1971)
14. Morf, W. E. et al.: Cation Selectivity of Neutral Macrocyclic and Nonmacrocyclic Complexing Agents in Membranes, in: Progr. Macrocyclic Chem., Vol. I (R. M. Izatt and J. J. Christensen (Eds.)), John Wiley and Sons, New York, 1979, p. 1
15. Diebler, H. et al.: Pure and Appl. Chem. *20*, 93 (1969)
16. Burgermeister, W., Winkler-Oswatitsch, R.: Topics in Curr. Chem., Vol. 69, Springer-Verlag, Berlin, 1977, p. 204
17. Pressman, B. C.: Ann. Rev. Biochem. *45*, 501 (1976)
18. Shemyakin, M. M. et al.: Experientia *27*, 548 (1965)
19. Ivanov, V. T. et al.: Biochem. Biophys. Res. Commun. *34*, 803 (1969)
20. Ovchinnikov, Yu. A., Ivanov, V. T., Shkrob, A. M.: The Chemistry and Membrane Activity of Peptide Ionophores, in: Molecular Mechanisms of Antibiotic Action on Protein Biosynthesis and Membranes (Munoz, E., Garcia-Ferrandiz, F. and Vazquez, P. (Eds.)), Elsevier, Amsterdam, 1972, p. 459
21. Lardy, H. A., Graven, S. N., Estrada-O, S.: Fed. Proc. *26*, 1355 (1967)
22. Ovchinnikov, Yu. A. et al.: Biochem. Biophys. Res. Commun. *37*, 668 (1969)
23. Popov, E. M. et al.: Chem. Natur. Prod., *616* (1970)
24. Dobler, M., Dunitz, J. D., Krajewski, J.: J. Mol. Biol. *42*, 603 (1969)
25. Beck, J. et al.: Helv. Chim. Acta *45*, 620 (1962)
25A. Dobler, M., Dunitz, J. D., Kilbourn, B. T.: Helv. Chim. Acta *52*, 2573 (1969)
26. Pedersen, C. J.: Org. Synth. *52*, 66 (1972)

27. Pederson, C. J.: Synthetic Multidentate Macrocyclic Compounds, in: Synthetic Multidentate Macrocyclic Compounds (R. M. Izatt and J. J. Christensen (Eds.)), Academic Press, New York, 1978, p. 10
28. Cram, D. J., Cram, J. M.: Science *183*, 803 (1974)
29. Newcomb, M. et al.: J. Amer. Chem. Soc. *101*, 4491 (1979)
30. Pressman, B. C.: Fed. Proc. *27*, 1283 (1968)
31. Westley, J. W.: Annu. Rep. Med. Chem. *10*, 246 (1975)
32. Agtarap, A. et al.: J. Amer. Chem. Soc. *89*, 5737 (1967)
33. Pinkerton, M., Steinrauf, L. K.: J. Mol. Biol. *49*, 533 (1970)
34. Steinrauf, L. K., Pinkerton, M., Chamberlin, J. W.: Biochem. Biophys. Res. Comm. *33*, 29 (1968)
35. Stemplel, A., Westley, J. W., Benz, W.: J. Antibiotics *22*, 384 (1969)
36. Lutz, W. K., Wipf, H. K., Simon, W.: Helv. Chim. Acta *53*, 1741 (1971)
37. Pressman, B. C., de Guzman, N. T.: Ann. N. Y. Acad. Sci. *264*: 373 (1975)
38. Richardson, L. F. et al.: J. Animal Sci. *43*, 657 (1976)
39. Feinman, S. E., Matheson, J. C.: Draft Environmental Impact Statement: Subtherapeutic Antibacterial Agents in Animal Feeds. Available from Hearing Clerk, Food and Drug Administration, Room 4-65, 5600 Fishers Lane, Rockville, Maryland 20857, 1978: pp. A100–A108
40. Grell, E., Funck, Th., Eggers, F.: Dynamic Properties and Membrane Activity of Ion Specific Antibiotics, in: Molecular Mechanisms of Antibiotic Action on Protein Biosynthesis and Membranes (E. Munoz, F. Garcia-Fernandez, D. Vazquez (Eds.)), Elsevier, Amsterdam, 1972, p. 647
41. Ohnishi, M., Urry, D. W.: Biochem. Biophys. Res. Commun. *36*, 194 (1969)
42. Bystrov, V. F. et al.: Eur. J. Biochem. *78*, 63 (1977)
43. Mayers, D. F., Urry, D. W.: J. Amer. Chem. Soc. *94*, 77 (1972)
44. Maigret, B., Pullman, B.: Theor. Chim. Acta *37*, 17 (1975)
45. Feigenson, G. W., Meers, P. R.: Nature *283*, 313 (1980)
46. Johnson, S. M. et al.: J. Amer. Chem. Soc. *92*, 4428 (1970)
47. Duncan, A. B. F., Matsen, F. A., in: Technique in Organic Chemistry (A. Weissberger (Ed.)), Interscience, New York, 1966, p. 581
48. Westley, J. W. et al.: J. Med. Chem. *16*, 397 (1973)
49. Haynes, D. H., Chiu, V. C. K., Watson, B.: Arch. Biochem. Biophys. *208*, 73 (1980)
50. Painter, G. R., Pressman, B. C.: Biochem. Biophys. Res. Commun. *91*, 1117 (1979)
51. Painter, G. R., Pressman, B. C.: Fed. Proc. *40*, 568 (1981)
52. Painter, G. R., Pressman, B. C.: Fed. Proc. *39*, 2157 (1980)
53. Reichardt, C.: Angew. Chem. *77*, 30 (1965)
54. Burgess, J.: Metal Ions in Solution, John Wiley and Sons, New York 1978
55. Degani, H., Friedman, H. L.: Biochem. *14*, 3755 (1975)
56. Wellman, K. M. et al.: J. Amer. Chem. Soc. *87*, 66 (1965)
57. Miles, P. W., Urry, D. W.: Biochem. *7*, 2791 (1968)
58. Painter, G. R., Pressman, B. C.: Biochem. Biophys. Res. Commun. *97*, 1268 (1980)
59. Urry, D. W.: Optical Rotation and Biomolecular Conformation, in: Spectroscopic Approaches to Biomolecular Conformation (D. W. Urry, Ed.), Amer. Med. Ass., Chicago, 1970, p. 34
60. Patel, D. J., Shen, C.: Proc. Natl. Acad. Aci. U.S. *73*, 1786 (1976)
61. Shen, C., Patel, D. J.: ibid. *74*, 4734 (1977)
62. Painter, G. R., Pressman, B. C.: The Conformational Dynamics of the Carboxylic Ionophore Lasalocid A Underlying Cation Complexation-Cecomplexation and Membrane Transport. Submitted to Biochemistry, May, 1981
63. Pullman, A., Pullman, B.: Quart. Rev. Biophysics *7*, 505 (1975)
64. Madison, V., Kopple, K. D.: J. Amer. Chem. Soc. *102*, 4855 (1980)
65. Huber, E., Warshel, A.: Chem. Phys. *6*, 436 (1974)
66. Rohrer, D.: Editmodel, in: PROPHET Molecules: A User's Guide to the Molecule Facilities of the PROPHET System (W. P. Rindone and Kushl, T. (Eds.)), p. 5 (31), Bolt Beranek and Newman Inc., Cambridge, 1980
67. Deber, C. M., Pfeiffer, D. R.: Biochem. *15*, 132 (1976)
68. Pfeiffer, D. R., Lardy, H. A.: Biochem. *15*, 935 (1976)
69. Moscowitz, A. et al.: J. Amer. Chem. Soc. *83*, 4461 (1961)

70. Djerassi, C. et al.: ibid. *84*, 870 (1962)
71. Moffitt, W. et al.: ibid. *83*, 4013 (1961)
72. Crabbé, P.: ORD and CD in Chemistry and Biochemistry, Academic Press, New York, 1972, p. 35
73. Kinashi, H. et al.: Studies Acta. Cryst. *B31*, 2411 (1975)
74. Haynes, D. H., Pressman, B. C.: J. Membrane Biol. *16*, 195 (1974)
75. Feynman, R. P., Leighton, R. B.: The Feynman Lectures on Physics, Vol. II, Addision-Wesley Pub. Co. Reading, Mass, 1964
76. Haynes, D. H., Staerk, H.: J. Membrane Biol. *17*, 313 (1974)
77. Haynes, D.: ibid. *17*, 341 (1974)
78. Caswell, A. H., Pressman, B. C.: Biochem. Biophys. Res. Commun. *49*, 292 (1972)
79. Pfeiffer, D. R., Reed, P. W., Lardy, H. A.: Biochem. *13*, 4007 (1974)
80. Kolber, M. A., Haynes, D. H.: Fluorescence Study of the Divalent Cation Transport Mechanism of A23187 in Phospholipid Membranes, Accepted for publication in Biophysical J., May, 1981
81. Ladbrooke, B. D., Chapman, D.: Chem. Phys. Lipids *3*, 304 (1969)
82. Jacobs, R. E., Hudson, B. S., Andersen, H. C.: Biochem. *16*, 4349 (1977)
83. Reed, P. W.: Meth. Enzymol. *55*, 435 (1979)
84. Pressman, B. C.: Fed. Proc. *32*, 1698 (1973)
85. Levy, J. V., Cohen, J. A., Inesi, G.: Nature *242*, 461 (1973)
86. Schwartz, A. et al.: Circ. Res. *34*, 102 (1974)
87. de Guzman, N. T., Pressman, B. C.: Circulation *49*, 1072 (1974)
88. Schaffer, S. W. et al.: Biochem. Pharmacol. *23*, 1609 (1974)
89. Holland, D. R., Steinberg, M. I., Armstrong, W.: Proc. Soc. Exp. Biol. Med. *148*, 1141 (1975)
90. Pressman, B. C., de Guzman, N. T.: Ann. N.Y. Acad. Sci. *227*, 380 (1974)
91. Pressman, B. C., de Guzman, N. T., Somani, P.: The Pharmacologist *17*, 245 (1975)
92. Pressman, B. C., Lattanzio, F. A.: Frontiers of Biol. Energetics *2*, 1245 (1978)
93. Douglas, W. W.: Secretory Mechanisms of Exocrine Glands, Munksgaard, Copenhagen, p. 116, 1974
94. Fahim, M., del Valle, G., Pressman, B. C.: Fed. Proc. *40*, 711 (1981)
95. Rubin, R. W., Corcoran, J., Pressman, B. C.: J. Cell Biol. *83*, 434 (1979)
96. Suchard, S. J., Rubin, R. W., Pressman, B. C.: J. Cell. Biol. *87*, 303a (1980)
97. Corcoran, J. J. et al.: Fed. Proc. *39*, 2102 (1980)
98. Harris, E. J., Pressman, B. C.: Nature *216*, 918 (1967)
99. Henderson, P. J. F., McGiven, J. P., Chappell, J. B.: Biochem. J. *111*, 521 (1969)
100. Pressman, B. C., Greenwald, R.: Biophys. J. *17*, 151a (1977)
101. Heeb, M. J., Pressman, B. C.: Permeability Studies on Erythrocyte Ghosts with Ionophorous Antibiotics, in: Molecular Mechanisms of Antibiotic Action on Protein Biosynthesis and Membranes (E. Munoz, F. Garcia-Fernandez and D. Vasquez (Eds.)), Elsevier, Amsterdam, 1972, p. 603
102. Krasne, S., Eisenman, G.: J. Membrane Biol. *30*, 1 (1976)
103. Ciani, S., Eisenman, G., Szabo, G.: J. Membrane Biol. *1*, 1 (1969)
104. Duesler, E. N., Paul, I. C.: The X-ray Structure of the Polyether Antibiotics, in: Polyether Antibiotics: Carboxylic Acid Ionophores, Vol. 2, (J. W. Westley (Ed.)), Marcel Dekker Inc., New York, 1982
105. Yamazaki, N. et al.: Tetrahedron Lett. *27*, 2429 (1978)

Bioorganic Modelling
Stereoselective Reactions with Chiral Neutral Ligand Complexes as Model Systems for Enzyme Catalysis

Richard M. Kellogg

Department of Organic Chemistry, University of Groningen, Nijenborgh 16,
9747 AG Groningen/Netherlands

Table of Contents

1 Introduction 111

2 Enantioselective Reactions 113

3 Cyclodextrins 115

4 Peptides 116

5 Catalytically Active Synthetic Macrocycles 118
   5.1 Crown Ethers — General Design Points 118
   5.2 Crown Ether Derived Compounds That Imitate Proteases 121
   5.3 Cyclic Hydrocarbon Systems that Imitate Proteases 125
   5.4 Macrocycles with Hydrogenase Activity 126
   5.5 Macrocycles Capable of Catalyzing Other Bond-Forming or
       Bond-Breaking Reactions 139

6 Summary 141

7 Acknowledgement 142

8 References 142

Richard M. Kellogg

1 Introduction

In 1926 Leopold Ruzicka, on the occasion of his inaugural address as newly appointed professor of organic chemistry at the University of Utrecht, observed

„Es ist also das Endziel der organisch-chemischen Forschung einen bestimmten Teil des Materials zu liefern für die Beantwortung einer die Menschheit tief bewegenden Frage: welches sind die naturwissenschaftlich klar faßbaren Grundlagen der Lebensvorgänge? Gar mannigfaltig sind die Wege und Möglichkeiten, die der chemischen Forschung zur Erreichung dieses Zieles zur Verfügung stehen. Es sind nicht nur jene wichtig, die direkt an das Hauptziel führen; auch scheinbar nebensächliche und fernliegende Einzelheiten können manchmal in unerwarteter Weise die große letzte Aufgabe fördern"[1)].

The words remain true. The years since 1926 have brought us, however, to a significant new stage. The point has now been reached that, armed with remarkable amounts of information on the structure and mechanism of action of large biomolecules, steps can now be contemplated to blend that knowledge with good understanding of synthetic and mechanistic chemistry into man-made compounds designed to carry out specific functions.

There are many directions in which this can proceed. This review, however, will be for reasons of subject matter and space restricted to a "state-of-the-art" discussion of only one aspect of such work. This aspect will be the bioorganic modelling of certain enzymic processes with fairly small molecules, naturally occurring or synthetic, which have the ability to complex the substrate in a rapid pre-equilibrium, just as in an enzyme. Because of the subject content of this book these compounds will be in almost all cases macrocycles and they will usually have also the capacity for the recognition of enantiomers of a potential substrate.

Most effort so far has been concentrated on the design of synthetic systems capable of mimicking the selectivity and/or speed of enzymic processes. A portion of such efforts will also be the subject of this chapter. However, a few generalizations should be made at this point. First, if the path to the design and synthesis of "artificial enzymes" is to be followed, blind and unthinking imitation of Nature should be avoided. Drastic but creative innovations dictated by synthetic access are in order. For example, it is not possible to synthesize all too readily a long polypeptide with a defined order of amino acids and having specified properties. Other approaches must be followed. At the current stage of synthetic knowledge the use of molecules far smaller than enzymes is virtually mandatory. Synthetic strategies with a degree of flexibility can often be developed for the preparation of covalently bonded organic molecules with molecular weights 10^{-3} to 10^{-4} those of typical enzymes. This means the loss of enormous amounts of structure as found in enzymes; it is a moot point, however, whether this structure represents redundant or necessary information in regard to the chemistry of the reaction to be catalyzed.

A further simplification often necessary is the use of organic solvents rather than water. However, the tendency that many enzymes have to develop active sites in hydrophobic pockets from which water is fairly well excluded provides some excuse for this pragmatic course of action [2)].

The consequences of using synthetic molecules far smaller than normal enzymes, assuming that the desired catalytic aspects can and have been built into the small molecule, will doubtlessly be felt both in speed and selectivity of action. However, one can usually tolerate great losses in reaction speed relative to an enzyme and still have a system that is chemically interesting and useful. Absolute selectivity with regard to substrate is also usually not wanted. Most often reaction of a *family* of structurally related compounds is desired and hence some tolerance in the "fit" is usually needed. What is required is a good recognition of the site at which a reaction should occur, if there is a choice of sites, and good chiral discrimination between enantiomeric substrates. The difficulties should not be underestimated of achieving a balance between these desirable aspects and what in reality is possible.

To achieve some of the desirable features mentioned above one piece of enzymic chemistry should be embodied, namely that the surface on which the catalyzed reaction occurs should be complementary (properly positioned hydrogen bonds, optimal orientation of charges, relief of strain, etc.) to the *transition state* of the reaction to be catalyzed. [3,4] This is doubtlessly one of the things available in the large amount of structure present in an enzyme perfected through evolution; optimal complementarity will be harder to achieve in a smaller synthetic structure. This problem of complementarity will be a major barrier in achieving rate enhancements approaching those of enzymes.

A final general point has to do with evolution. Many biochemically catalyzed processes have an important aspect of spontaneity in that a good part, if not all, of the chemistry will also proceed non-enzymatically. The remarkably spontaneous aspects of the cyclization of squalene epoxide to lanesterol come to mind [5] as does the stereochemically less complex but just as spectacular tropinone (1) synthesis of Robinson (eq. 1) [6].

$$HC(CH_2)_2CH + {}^{\ominus}O_2CCH_2\overset{O}{C}CH_2CO_2^{\ominus} \cdot Ca^{2\oplus} + CH_3NH_2 \longrightarrow \mathbf{1} + CO_2 \quad (1)$$

The point is that some enzymic processes may operate with active site components that by themselves are fairly capable for carrying out the desired reaction. These components have become incorporated into a peptide chain forming an enzyme that has become very efficient in the course of evolution. A detailed discussion of the association of chemistry and evolution has been given by Visser. [7] Such considerations are especially important with coenzymes, these being particularly loved subjects for model study (and will be discussed again later in this review). The point to be drawn from this is that bioorganic modelling of enzymic chemistry is easier if the chemistry tends to proceed spontaneously. Although this sounds like a statement of the obvious, confusion leading to poor design of experiments has occurred with certain coenzymes, for example biotin, which has consistently been assigned an activating role that it does not fulfil. [7a,b,8] These evolutionary points will be discussed where appropriate in the text.

2 Enantioselective Reactions

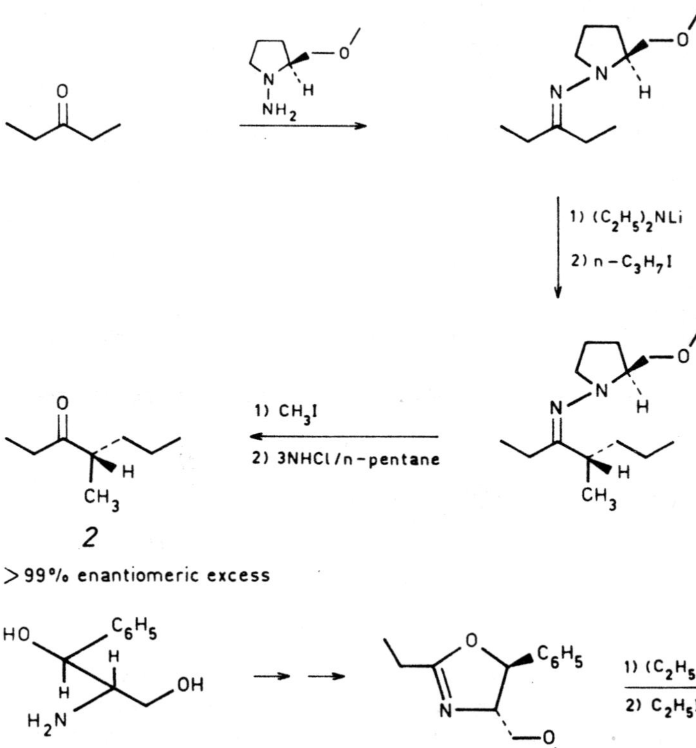

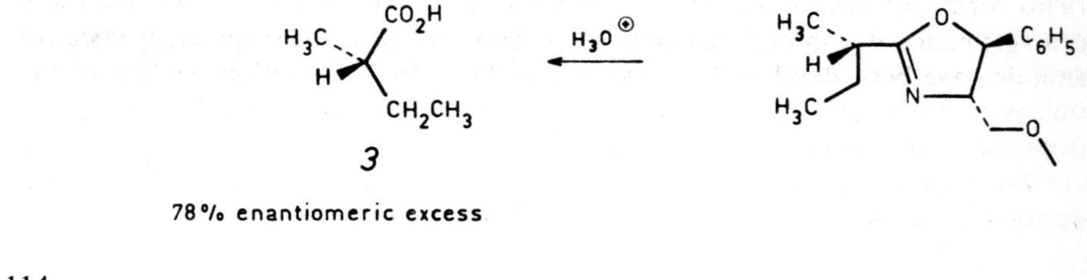

A major consideration of this review will be how to achieve enantioselective reactions using macrocyclic systems. Macrocycles are attractive for this purpose because of the possibility of defining well the structure of the complex leading to reaction. However, there is no magic in macrocycles; highly enantioselective syntheses using well-designed non-macrocyclic systems have been developed in recent years. Four pertinent examples are shown in eqs. 2–5. The success of these reactions lies undoubtedly for a good part in chelation, which enforces rigidity during the transition state for reaction. The most important point to be learned is that even with small molecules that good to excellent enantioselectivities can be achieved in reactions ranging from

(2)

2

>99% enantiomeric excess

(3)

3

78% enantiomeric excess

(4)

4

62% enantiomeric excess

(5)

5

95% enantiomeric excess

carbon-carbon bond formation (*2, 3*) [9,10] to carbon-sulfur bond formation (*4*) [11] to epoxidation (*5*) [12] of allylic alcohols by making use of only relatively small groups without extremely large steric demands. The free energy differences between the diastereometric transition states are not large — about 2 kcal/mol for an 80% enantiomeric excess [10a] — which illustrates how much can be accomplished with the cumulative effects of different small interactions.

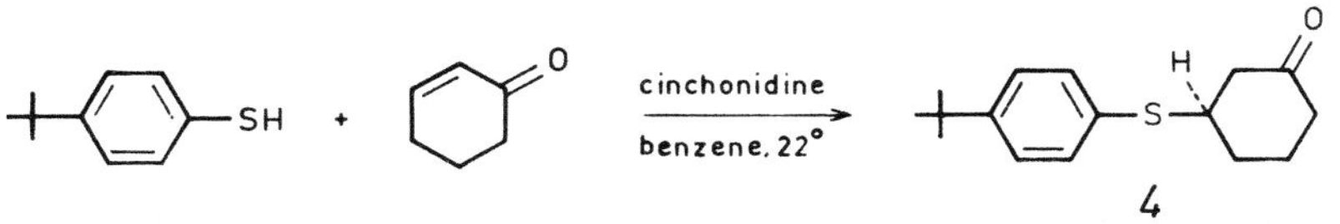

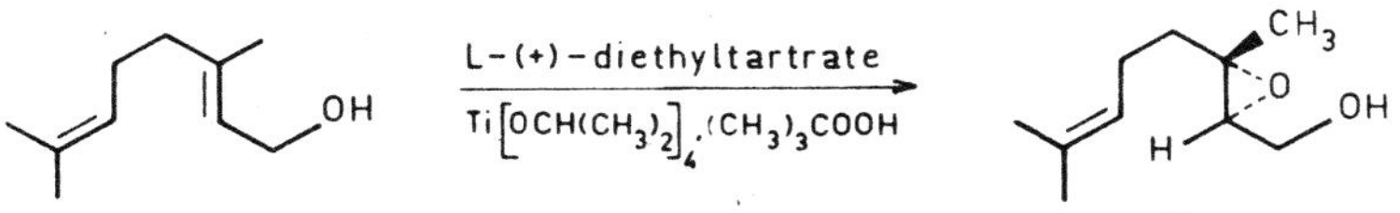

3 Cyclodextrins

Cyclodextrins are cyclic glucose oligomers (*6a*) having the shape roughly depicted in (*6b*), as a cylindrical form with the primary hydroxyl group at the more restricted end of the cylinder (most of the hydroxyl groups have been omitted from the drawing).

6a *6b*

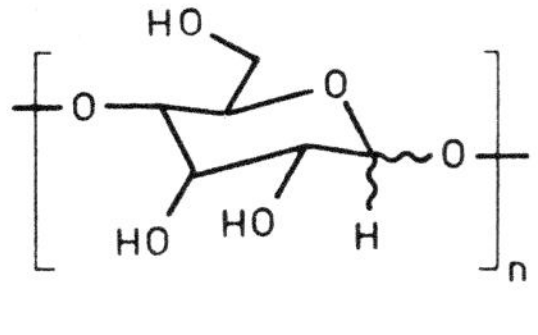

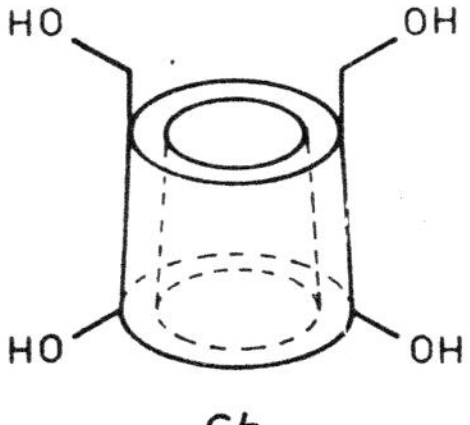

The interior is relatively apolar relative to water and is large enough (internal diameter about 4.5 Å and depth about 6.7 Å for α-cyclodextrin, which has six α-linked glucose units and about 7.5 Å internal diameter for β-cyclodextrin, typesetting error which is a β-linked glucose heptamer [13] to accommodate apolar guests of appropriate size such as benzene derivatives.

This forms an ideal basis for the synthesis of "enzyme models" since naturally occurring material with built-in complexing ability can be used. Surprisingly effective methods have been developed for the reasonably selective functionalization of the primary hydroxyl groups allowing the incorporation of various catalytically active groups. In some cases recognition of enantiomers or enantioselective syntheses have been involved. [14] All this work has been competently and extensively reviewed [15] and hence need not be rediscussed here.

There are definite limitations to the cyclodextrins. The presence of many hydroxyl groups makes the problem of selective functionalization difficult. Moreover, the glycoside linkages are only stable in neutral or basic solution, thereby restricting somewhat the chemical reactions than can be studied.

4 Peptides

Although not macrocyclic structures, small peptides, not enzymes, used to catalyze organic reactions, should be briefly mentioned in the context of this chapter. Interest has centered for the greater part on hydrolytic reactions related to those catalyzed by, for example, chymotrypsin. [16] Such studies usually involve the use of activated esters, usually p-nitrophenolates (7), which hydrolyze at readily followable rates under weakly basic conditions or in the presence of various nucleophiles.

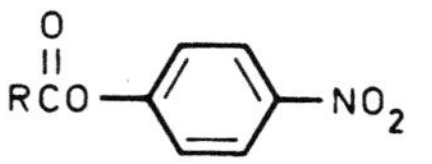

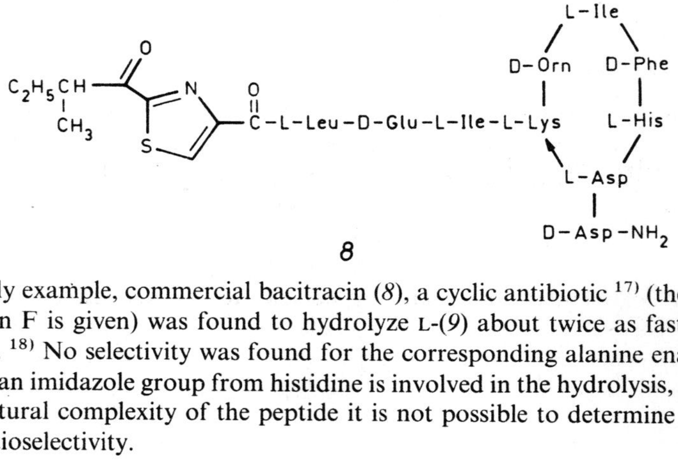

In an early example, commercial bacitracin (8), a cyclic antibiotic [17] (the structure of bacitracin F is given) was found to hydrolyze L-(9) about twice as fast as the D-enantiomer. [18] No selectivity was found for the corresponding alanine enantiomers. Most likely an imidazole group from histidine is involved in the hydrolysis, but owing to the structural complexity of the peptide it is not possible to determine the origin of the enantioselectivity.

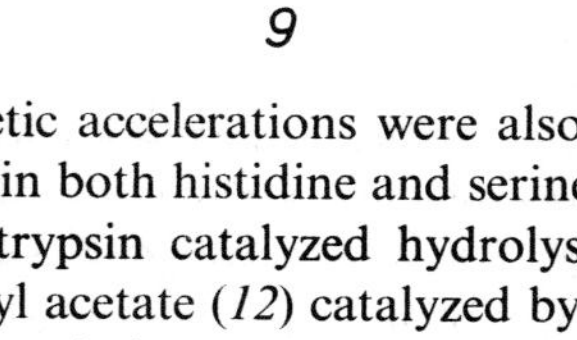

Mild kinetic accelerations were also found with the pentapeptides (10) and (11), which contain both histidine and serine, both of which are involved at the active site in β-chymotrypsin catalyzed hydrolyses. The catalytic constant for hydrolysis of p-nitrophenyl acetate (12) catalyzed by (10) is almost 16 times that of imidazole and for (11) the catalytic constant is 24.5 times greater. [19] However, the modestness of the catalytic effects is brought home on realizing that the catalytic constant for hydrolysis of (12) by α-chymotrypsin is 10^4 that of imidazole. With (11) fairly good chiral recognition was found for hydrolysis of the enantiomers of (13), the catalytic constants being 26 times greater than that for histidine for the D-enantiomer of (13) and 19 times greater for the L-enantiomer. Again, however, it is not possible to pinpoint the structural basis for these differences in hydrolysis rates.

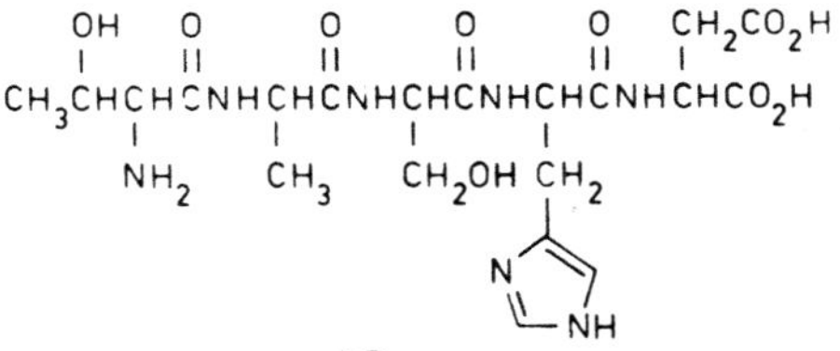

10

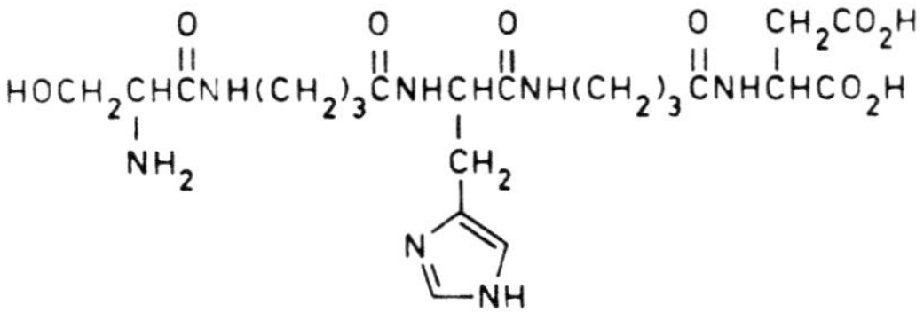

11

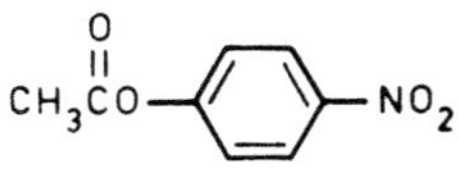

12

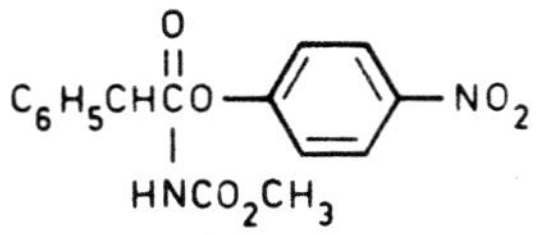

13

Naturally occurring peptides can also be used to catalyze reactions other than hydrolytic. Although the present survey is not exhaustive one can note, for example, that in the epoxidation of α,β-unsaturated ketones by basic hydrogen peroxide (eq. 6) that the polypeptide (14) acts as an efficient catalyst producing enantiomeric excesses of (15) of up to 93% [20].

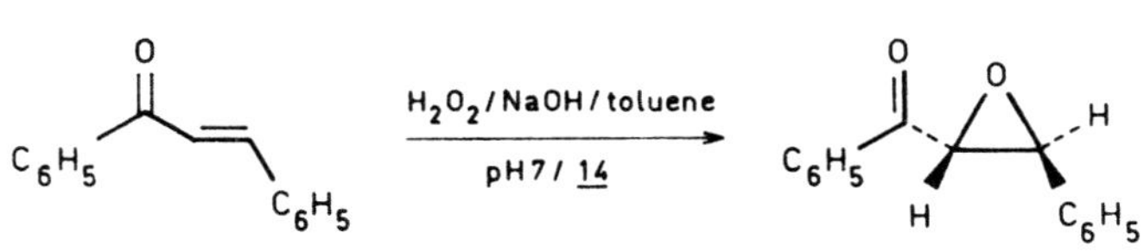

(6)

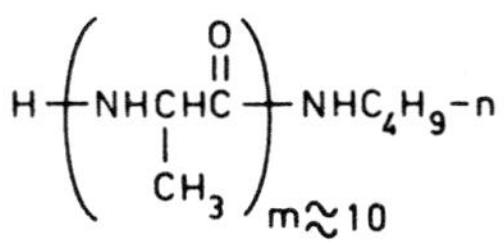

14

With a larger peptide, bovine serum albumin, molecular weight 66,000, which has three binding sites per molecule, both reductions and oxidations have been catalyzed. In the presence of stoichiometric amounts of bovine serum albumin acetophenone is reduced by sodium borohydride to 1-phenylethanol in up to 78% enantiomeric excess (eq. 7). [21] The acetophenone is firmly complexed to the peptide during reaction.

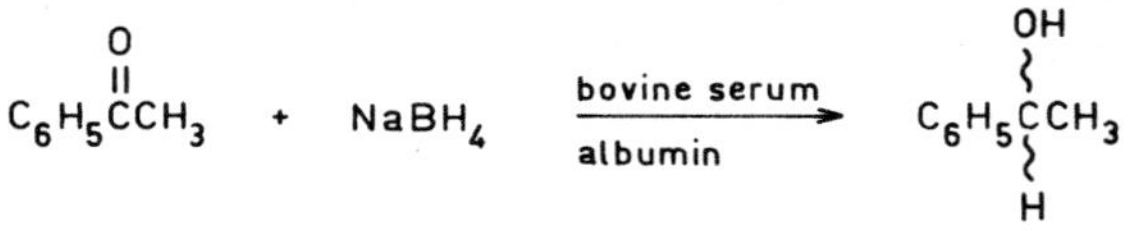

$$C_6H_5\underset{\underset{O}{\|}}{C}CH_3 + NaBH_4 \xrightarrow{\text{bovine serum albumin}} C_6H_5\underset{\underset{H}{|}}{\overset{\overset{OH}{|}}{C}}CH_3 \quad (7)$$

In a similar fashion sulfide (16) is oxidized to the sulfoxide (17) by $NaIO_4$ again in the presence of a stoichiometric amount of bovine serum albumin (eq. 8) in an enantiomeric excess of 81%. [22] Somewhat similar experiments have been reported

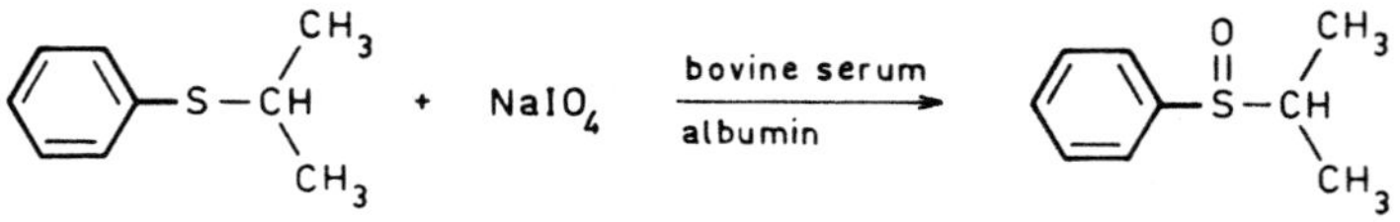

(8)

16   **17**

using kinetic resolution in the oxidation of a sulfoxide to an (achiral) sulfone. [22] In these experiments with bovine serum albumin there is again no structural basis for understanding the enantioselectivities observed. Nor are the reactions very general, at least at this stage of development.

5 Catalytically Active Synthetic Macrocycles

5.1 Crown Ethers — General Design Points

Following the development of effective synthetic routes to macrocyclic crown ethers as exemplified in eq. 9 for the synthesis of [18]crown-6 (18), extensive studies were initiated on the factors affecting their complexing ability. [23] The success of this work or, better said, the opportunity to carry it out depended on the availability of synthetic routes chiefly along the "template" lines depicted. [24] The synthetic opportunity to introduce significant structural modifications in a rational way is the aspect that lends the greatest advantage to the crown ethers. This synthetic flexibility is more difficult to

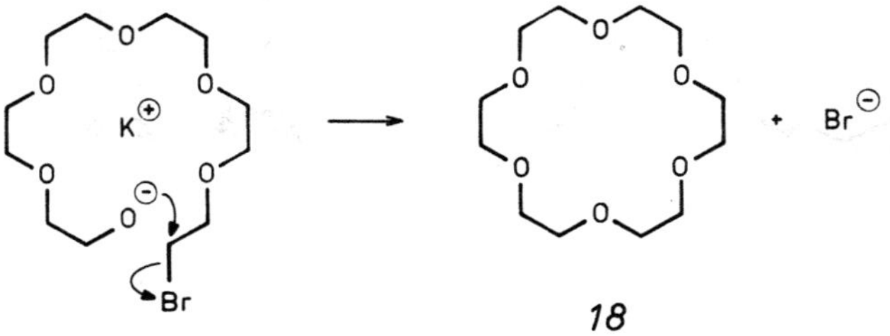

(9)

18

achieve with the cyclodextrins or, for example, with macrolide antibiotics, which have excellent complexing properties but usually so many functional groups that selective transformations on one particular substituent are not in any practical manner feasible (coupled often with scarcity of materials).

The classical model for binding of an *ion* in a crown ether involves three point bonding as illustrated for the ammonium ion complex with [18]crown-6 in *18-NH$_4^\oplus$a*, more easily drawn and visualized in *18-NH$_4^\oplus$b*, viewed from the top. [25] The most logical method of designing crown ethers capable of mimicking enzymic chemistry

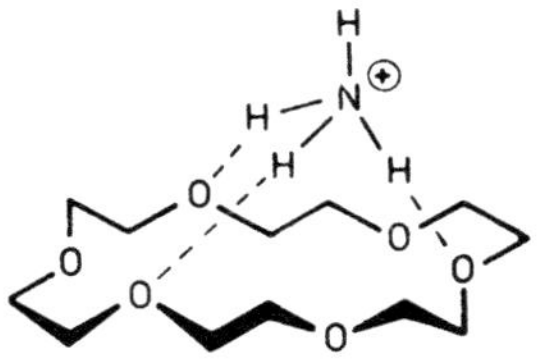

18a-NH$_4^\oplus$ 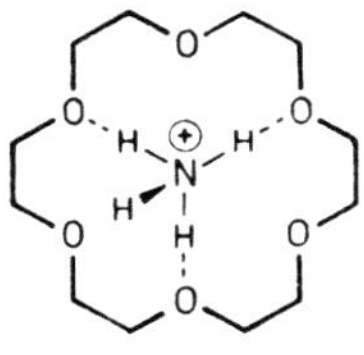   *18b*-NH$_4^\oplus$

is to assume the validity of the three point bonding model, use as guest an ammonium ion with a group R on which a reaction is to be carried out and then modify as desired the periphery of the crown ether.

The last aspect, modification of the periphery of the crown ether will be considered first. To the present time three main routes have been followed with as goal the synthesis of a *chiral* crown ether having also catalytically active groups in the periphery. Cram has brilliantly used bis-β-naphthol (*19*), which can be readily resolved into its enantiomers, as chiral component in the synthesis of crown ethers, an example being *20* (*a* and *b* being different projections) prepared using the potassium salt of the bis-phenolate of *19* and the bis-tosylate of pentaethyleneglycol. [26] The twisted nature of the bisnaphthyl linkage imparts a strong chiral bias to the system. A moderate price in complexing ability is paid for the disruption of the classical crown

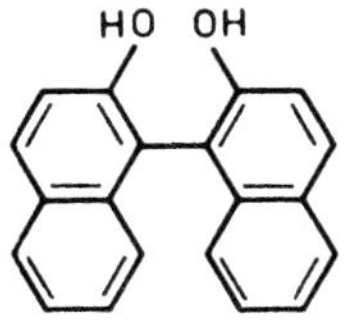

19

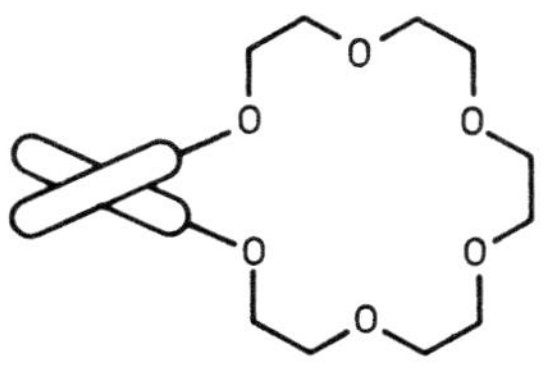

S-*20a* 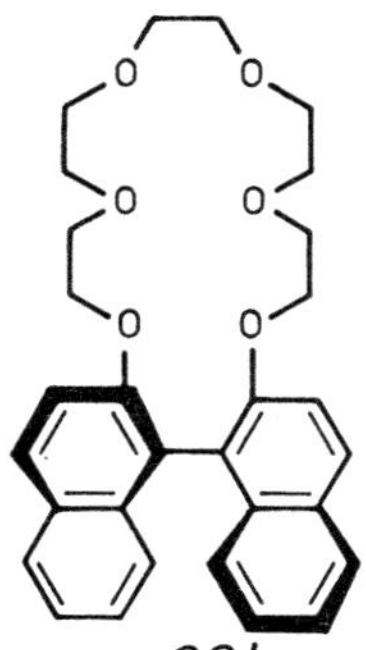   S-*20b*

ether system as well as the introduction of two *phenolate* oxygens of lowered basicity. The synthetic key to introduction of extra "arms", which can bear catalytically active groups, is a double Mannich reaction of (*19*) followed by acetylation and hydrolysis with LiAlH$_4$ to afford the *bis* (hydromethyl) derivative (*21*, eq. 10) which can be resolved easily [27].

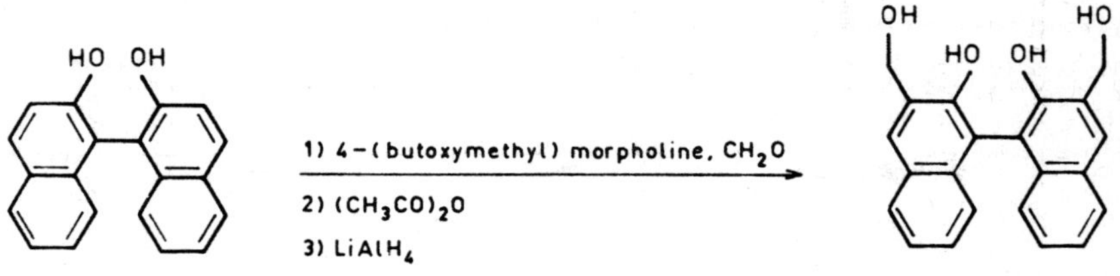

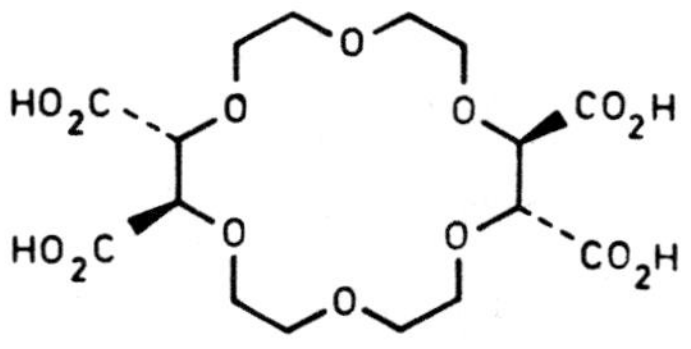

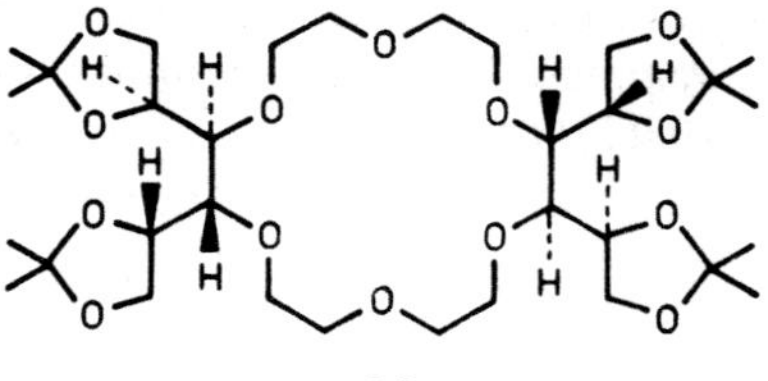

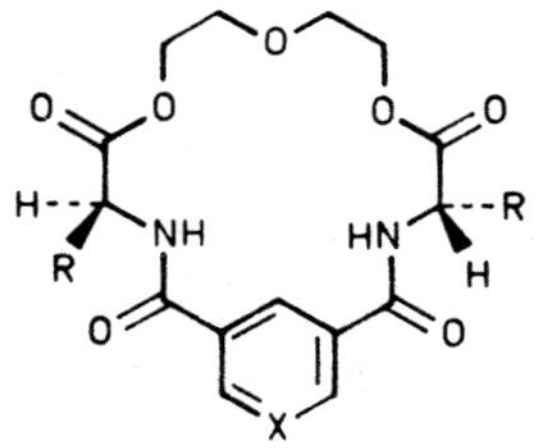

Another approach developed by the group of Lehn [28] requires a less drastic distortion of the periphery of the crown ether. For example in (*22*) derived from L-tartaric acid an [18]crown-6 periphery is still intact and four sites for attachment of functional groups are present. The D-enantiomer of tartaric acid is also available extending the synthetic flexibility derived by manipulation of the substituents [29].

Still another approach to chiral crown ethers that have potential applications for catalytic reactions is through incorporation of sugar residues as pioneered chiefly by Stoddart's group. [30] An example of one of the many compounds of this sort that has been synthesized (again by application of "templated" Williamson reactions) is (23) obtained from D-mannitol.

In our work we have developed methods to build heavily modified crown-like systems in which the chiral components are amino acids. [31] The general structural type is illustrated by (24a, b). In this case the macrocyclic crown ether system has been badly broken by extra substituents and heteroatoms (amide nitrogen, ester ether oxygen) of lowered basicity.

The chemical applications in catalytic reactions of these and related systems will be discussed in the following sections. One point of design that should be emphasized before proceeding on to the chemistry is that (20) and (22–24) have C_2 symmetry axes, which makes the faces of the system homotropic. Although this design feature is not mandatory, interpretations of complexing and reactions are greatly simplified if both faces of the reactive macrocycle are identical.

5.2 Crown Ether Derived Compounds That Imitate Proteases

Proteases like papain [32] have a cysteine sulfur as active nucleophile [33] and bind the substrate in a cleft. By using a crown ether ring instead of a cleft for binding, taking as substrate an ammonium salt that complexes via three-point bonding to oxygens of the crown ether, and introducing sulfur nucleophiles at the proper position to attack an activated carbonyl group in the ammonium substrate (ammonium salts of activated amino acid esters (25) — as illustrated in eq. 11 — are ideal) one has the potential for creating a synthetic protease, or at least a synthetic transacylase.

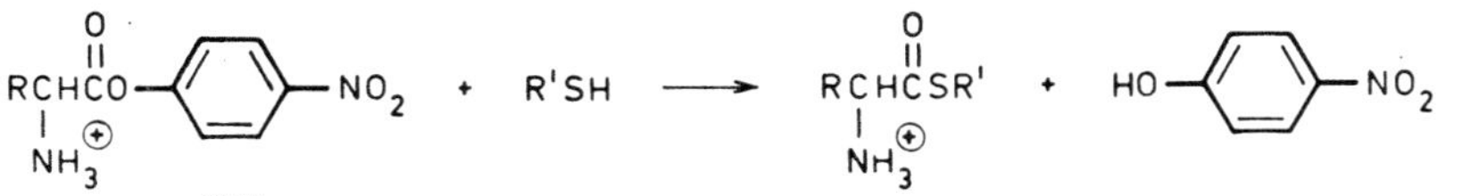

$$RCHCO\text{-}C_6H_4\text{-}NO_2 + R'SH \longrightarrow RCHCSR' + HO\text{-}C_6H_4\text{-}NO_2 \quad (11)$$
$$\underset{25}{NH_3^+} \qquad\qquad NH_3^+$$

The best examined system of this type is (26a, S-enantiomer illustrated) prepared and studied by Cram and coworkers. [26] The dimethyl derivatives (26b) had already been

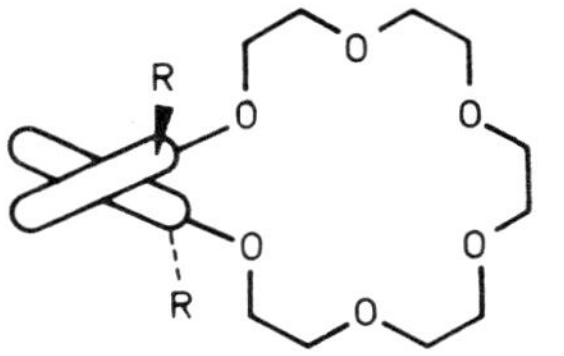

26
a) R=CH$_2$SH
b) R=CH$_3$

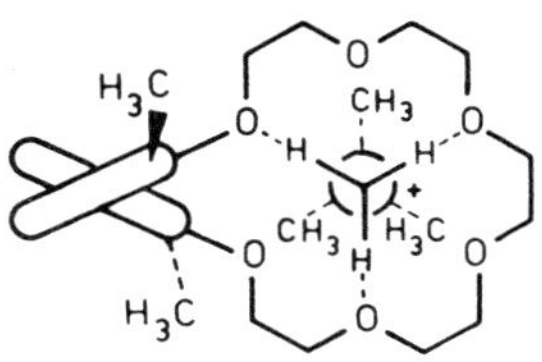

26b -(CH$_3$)$_3$CNH$_3^+$

established to bind $(CH_3)_3CNH_3^{\oplus}$, picrate$^{\ominus}$ in $CHCl_3$ at 25° with $\Delta G° = -6.4$ kcal/mol by means of the anticipated three-point binding ($26b$-$(CH_3)_3CNH_3^{\oplus}$, picrate$^{\ominus}$). Similar binding is expected with $26a$ with CH_2SH side arms.

In separate experiments with $HOCH_2CH_2SH$ as the thiol it was established that, at least at low concentrations, the thiolate anion, $HOCH_2CH_2S^{\ominus}$, is the only kinetically significant nucleophile in the thiolysis of 25 and that there are no appreciable buffer effects in the organic solvent mixtures used. The measured rates, determined by the appearance of p-nitrophenol or -phenolate depending on the pH are, however, those of transacylation (eq. 11) rather than subsequent (slow) solvolysis (eq. 12).

$$RCHCSR' + H-Osol \longrightarrow RCO-sol + R'SH \quad (12)$$
$$\underset{NH_3^{\oplus}}{|}$$

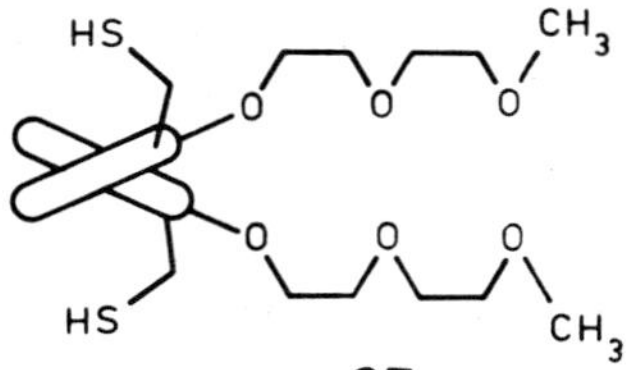

27

The effect of incorporating a cyclic structure is best evaluated by comparison with the transacylation rate constants for $26a$ with those for "open" analog (27) (S-enantiomer shown). For the transacylations of p-nitrophenolate esters of amino acid salts in 20% C_2H_5OH/CH_2Cl_2 at 25°, assuming no effective difference in pK's between ($26a$) and (27), cyclic $26a$ reacted consistently faster then "open" (27). For L-$25a$, R = $(CH_3)_2CHCH_2-$, $k_{26a}/k_{27} = 1170$, and for ($25b$), R = $(CH_3)_2CH-$, $k_{26a}/k_{27} = 160$, and for ($25a$), R = $C_6H_5CH_2-$, $k_{26a}/k_{27} = 490$ as typical values for the observed rate differences.

Not only is there a rate acceleration to be derived from the cyclic structure of ($26a$) but there is also structural recognition for the transacylation rates fall off with increasing size of the R group in (25). The rate for ($25c$, R = $(CH_3)_2CH$) is consistently lower than for ($25d$, R = CH_3), rate differences of 30–300 being observed depending on the series investigated. These effects probably arise from the fact that the tetrahedral intermediate from (S-$26a$) with (R-25) (or R-$26a$ with D-25) is more

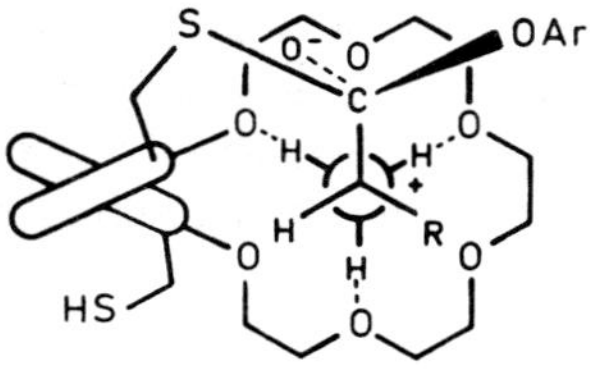

S-$26a$ + L-25
(more stable)

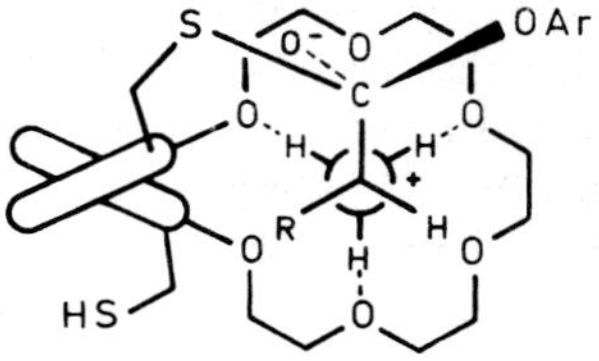

S-$26a$ + D-25
(less stable)

complementary to the macrocyclic structure than the tetrahedral intermediate formed from (R-*26a*) and D-*25* (S-*26a* with L-*25*). Note, of course, in the drawings that only the sterically most likely diastereomers of the tetrahedral intermediates are indicated. From CPK models the fit of the tetrahedral intermediates to the topology of the macrocycle appears to be reasonable but certainly not perfect.

From this work one can derive high hopes for the future because by *rational* design of complexes and before the experiment analyses of steric interactions (using CPK models) a synthetic catalyst was designed in a rational manner. The rate accelerations must of course to be made better and the problem of catalytic turnover has not yet been solved. In the reactions described the "catalyst" is actually used in about 50-fold excess. Solutions for these problems are challenges for the future.

Crown ethers based on tartaric acid (see *22*) have been used for similar transacylations.[34] The *tetra*-cysteine derivative (*28*) serves as a papain model. Because (*28*) has an undistorted [18]crown-6 periphery quite good complexation of

<pre>
              CH₂SH   CH₂SH
               | |
        CH₃O₂CCHNHOC ·.   O   O . CONHCHCO₂CH₃
                          ⌒ O ⌒
                         /   \
                        O   O
                         \   /
                          ⌒ O ⌒
        CH₃O₂CCHNHOC   O   O   CONHCHCO₂CH₃
               | |
              CH₂SH   CH₂SH

                      28 (L-cysteine derivative)

                    O   O
                    ‖   ‖
          ⊕
        H₃NCH₂CNHCHCO₂—⟨   ⟩—NO₂
                     |
                   CH₂C₆H₅
                      29
</pre>

ammonium ions is expected. Transacylations of p-nitrophenol esters of amino acid hydrobromides are indeed observed. Owing to the length of the cystinyl substituents the activated ester group of the substrate must be well separated from the ammonium ion center. This is achieved readily with dipeptides as substrates, an example being the p-nitrophenolate ester of glycinyl-L-phenylalanyl hydrobromide (*29*), which prior to thiolysis is probably bound to (*28*) as shown. In $CH_2Cl_2/CH_3OH/H_2O$ (97.9/2/0.1 ratio) the L-enantiomer of (*29*) undergoes acyl transfer to (*28*) about 50 times more rapidly than the D-enantiomer of (*29*). (L-*29*) appears to fit better in the cavity of (*28*). A detailed analysis of steric effects is difficult, however. Thiolysis definitely occurs in and not outside of the complex as established by the inhibitory effect of KBr, which competes for a binding site in (*28*). Benzylation of the thiol residues removes the catalytic activity of (*28*) entirely.

In (*28*), as with bis-naphthyl system (*26*), there is no turnover because the "catalyst" is used in excess in order to obtain rate data. Also both for (*26*) and (*28*) it is unfortunate that the thiol groups in the complexes are not all useable. In (*28*) the conformational flexibility of the cysteinyl arm is likely detrimental to the efficiency of catalysis

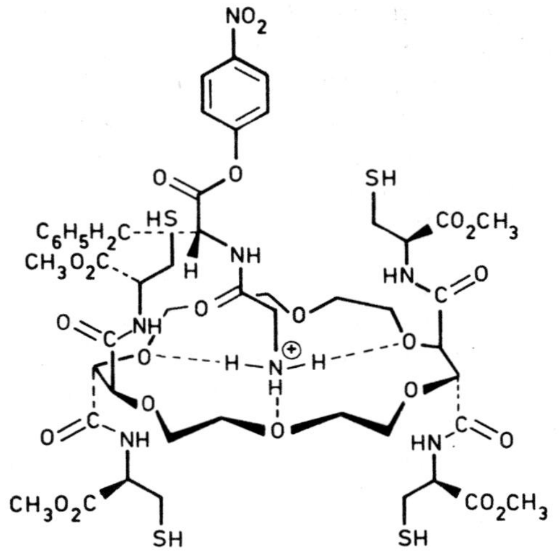

Richard M. Kellogg

28 – 29⊕

although the good complexing ability and high amino acid content does give it "enzyme-like" character.

Thiol catalyzed transacylations using the principles discussed above have also been investigated using crown ether systems *(30a–c)* *(31)* and *(32)*. [35, 36] The latter two compounds have particularly large chiral groups built into the periphery. The kinetic results with these compounds parallel those for *(26a)* and *(28)* with reasonable kinetic accelerations being found for transacylations. Chiral discriminations are modest, not exceeding a factor of 2 for D- and L-forms of *(25)*. Very likely too much conformational flexibility is still present in these compounds for highly efficient chiral recognition.

We have developed synthetic routes to resorcinol-based crown ethers like *(33)* [37]. These have been established to complex ammonium salts and to undergo conformational changes wherein the phenyl ring acts as a "hinge" moving the catalytically

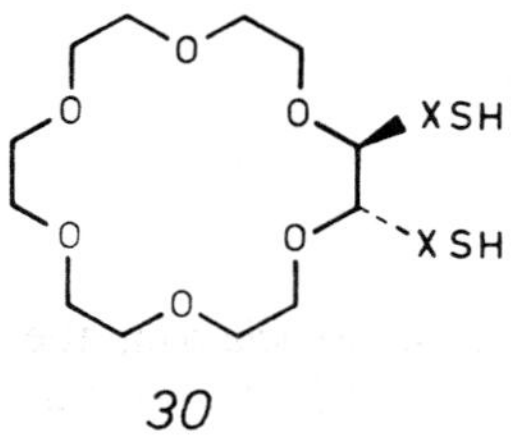

30

a) X = -CH$_2$-
b) X = -(CH$_2$)$_3$-
c) X = -CH$_2$O(CH$_2$)$_2$-

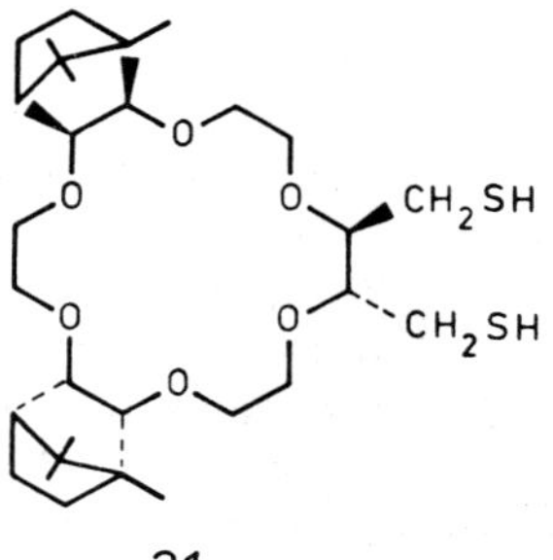

31

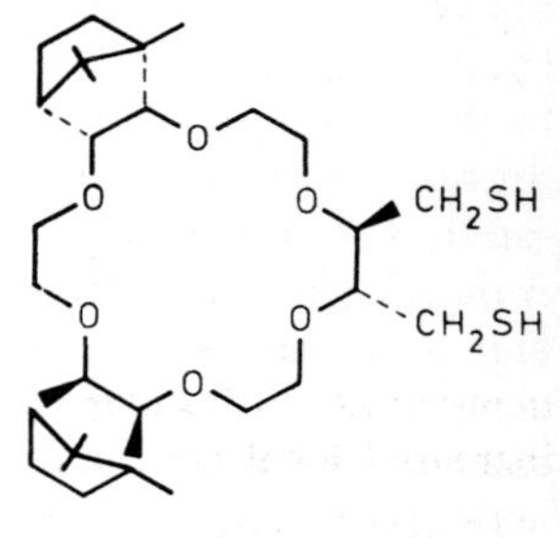

32

active thiol group back and forth relative to the macrocyclic ring. Transacylations with these compounds are being investigated.

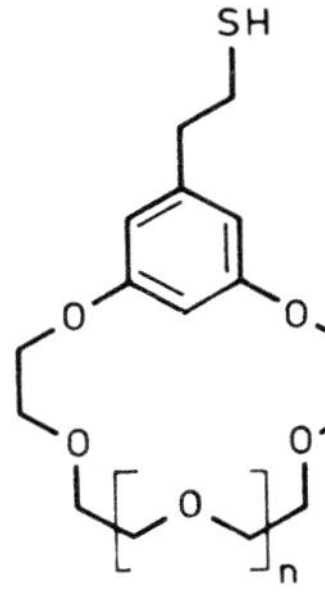

33

a) n = 2
b) n = 3

5.3 Cyclic Hydrocarbon Systems that Imitate Proteases

Crown ether derived systems have as attractive feature a good binding cavity for *ions*; the structural mixture of polar heteroatoms and connecting apolar hydrocarbon backbone also usually ensures a good solubility of the crown ether and its complexes in polar solvents.

There are, however, other binding forces that might be used to hold a, for example, *nonionic* substrate in a cavity. One such possibility that has been investigated is the use of "hydrophobic interactions". [38] This can be illustrated with an example. The macrocycles (*34*) and (*35*) [39], obtainable by means of acyloin condensation, followed by conversion to the oxime, possess functionalities that are potential catalysts for the hydrolysis of activated esters. [16] If hydrophobic interactions are important, an activated ester of a long chain fatty (i.e. hydrophobic) acid, for example, the p-nitrophenolate ester of lauric acid, should associate *in water* with the hydrophobic

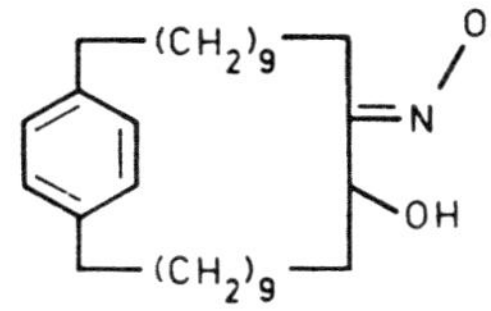

34 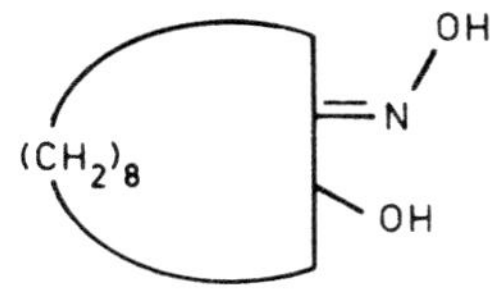   *35*

macrocycle. Although there is no control over the geometry of association, the increased local concentrations could produce rate enhancements. Indeed (*34*), but not (*35*), induces mild rate enhancements and the rate is inhibited by $Cu^{2\oplus}$ ions, which compete for the α-hydroxyoxime ligand.

A complication with such work, however, is the pronounced and expected tendency of such poorly soluble compounds to form micelles in water. [40] This tendency was underestimated in earlier work [41] on synthetic systems suggested to be models for esterases. To avoid micelle formation very low concentrations, usually $<10^{-6}$ M, must be used. The limits severely the range of practical applications.

A structurally more sophisticated hydrocarbon derived macrocycle is (36),[42] which also exhibits hydrolytic activity. At low concentrations (36), relative to imidazole alone, produces enhanced rates of hydrolysis of p-nitrophenyl esters of long chain acids. The suggestion that binding of substrate to catalyst occurs is supported by the fact that (37), which bears no catalytically active groups, *inhibits* the hydrolysis rate of p-nitrophenyl acetate by binding the substrate in the hydrophic cavity [43].

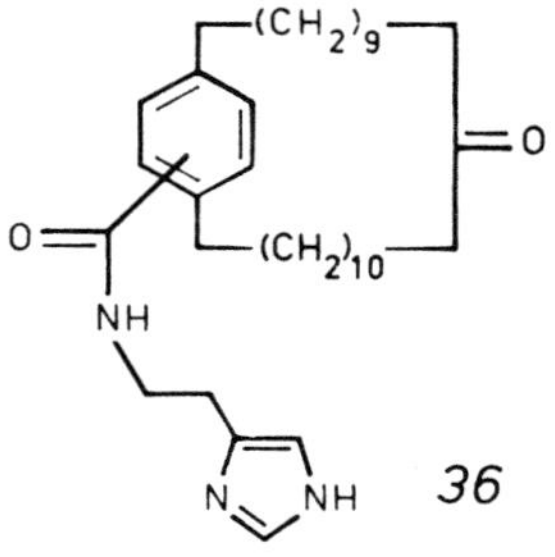

5.4 Macrocycles with Hydrogenase Activity

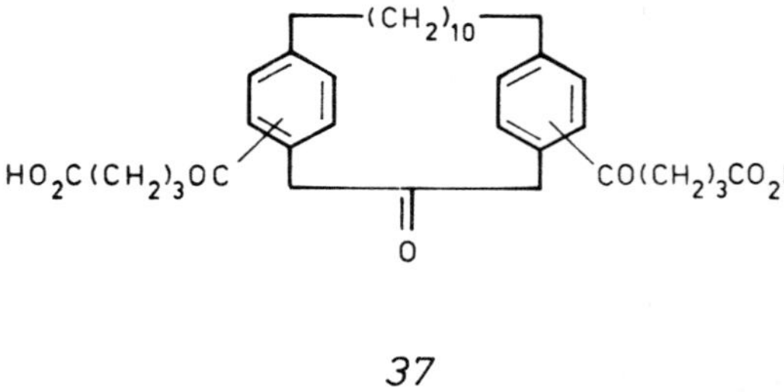

Excluding hydrolytic enzymes, about 70% of known enzymes require a cofactor and this figure becomes even higher if metal ions are included as cofactors. [44] An obvious chemical extrapolation from this information is to use the small coenzyme, or a similar molecule with the same chemistry, and to replace the accompanying protein (apoenzyme) and its function with synthetic material. To do this it is necessary that the chemical transformations desired be intrinsic to the coenzyme itself, the apoenzyme providing a binding site and carrying out an activating and regulatory role. As already mentioned coenzymes that are evolutionarily old are most likely to have the intrinsic chemistry desired.

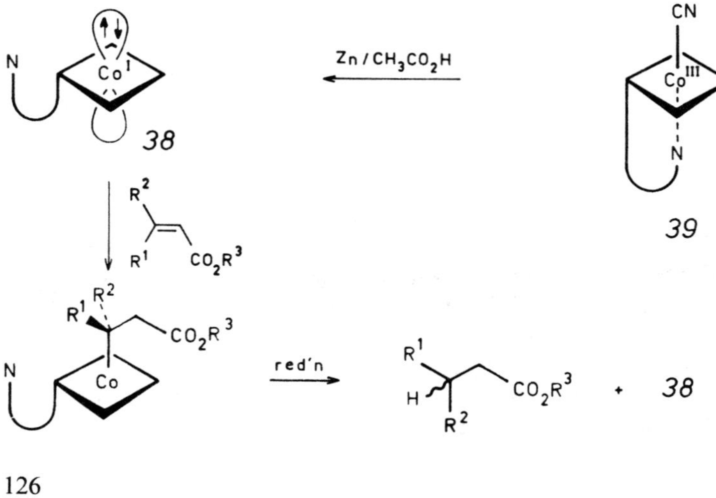

(13)

A case that may be mentioned briefly in the present context is that of cob(I)alamin (*38*) generated chemically from cyanocob(III)alamin (*39*). In extensive studies Fischli [45-49] has demonstrated that good degrees of enantioselectivity in *nonenzymic* reactions can be achieved as illustrated for a generalized hydrogenation scheme in eq. 13, in which the schematic drawings are intended to represent the corrin ring (perspective four-ring) and the attached benzimidazole in the "off" position.

This intrinsic reactivity needed for spontaneous reaction is certainly not present in every coenzyme, however. An example is biotine (*40*), which in its carboxylated form (*41*, eq. 14) is a form of *deactivated* rather than "activated CO_2". Activation by the accompanying apoenzyme is essential for CO_2 transfer. [8] This process has been discussed in terms of evolution [7a].

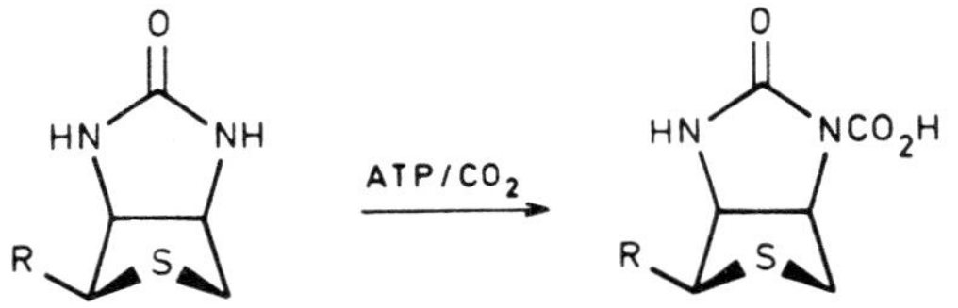

$$40 \xrightarrow{ATP/CO_2} 41 \tag{14}$$

Nicotinamide adenine dinucleotide (*42*) in the 1,4-dihydro form is a coenzyme that does lend itself well for use in model systems. The reactive portion is the 1,4-dihydronicotinamide ring (*43*), which is capable of acting as "hydride" donor towards many substrates; the pyridinium salt (*44*) acts in turn as "hydride" acceptor (eq. 15). The sugar nucleotide tail can be deleted without fatal consequences for the reactivity although the 1,4-dihydropyridine form (*44*) by itself has little tendency to reduce anything except the most reactive of potential substrates. [50] The problem is not

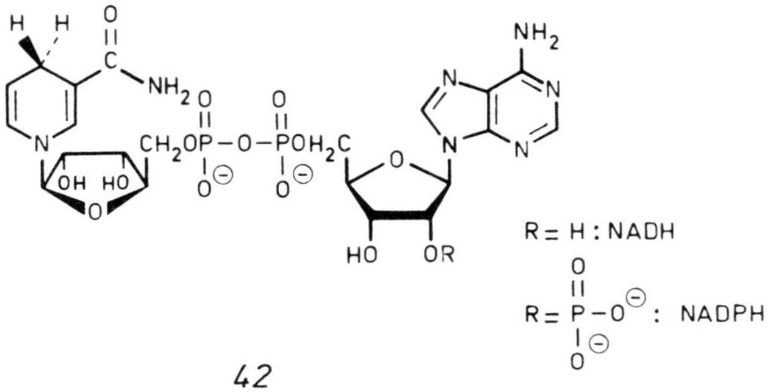

R = H : NADH

R = $P-O^\ominus$: NADPH

$$43 \xrightarrow[+ "H^\ominus"]{- "H^\ominus"} 44 \tag{15}$$

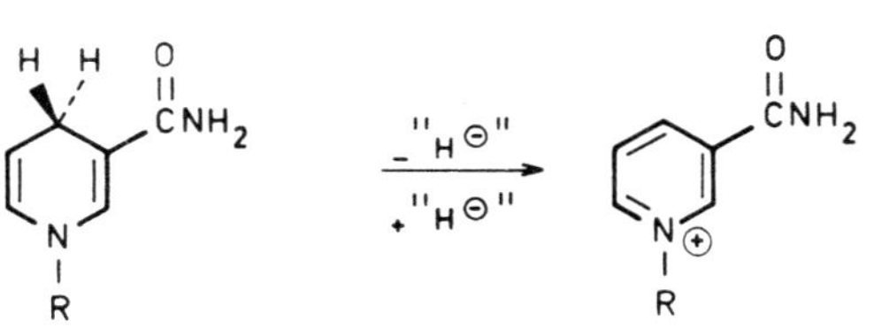

intrinsic lack of reduction potential of synthetic 1,4-dihydropyridines (N-benzyl-1,4-dihydronicotinamide (*43*, $R=CH_2C_6H_5$) has $E_0 = -361$ mV in nonaqueous solution compared to $E_0 = -315$ mV for NADH in aqueous solution) [51].

For reduction of carbonyl compounds *catalysis* is necessary. One of the major contributing effects to *enzymic* catalysis is polarization of the carbonyl substrate by an electrophile (eq. 16), which is usually imidazole or a zinc ion embedded in the peptide chain [52].

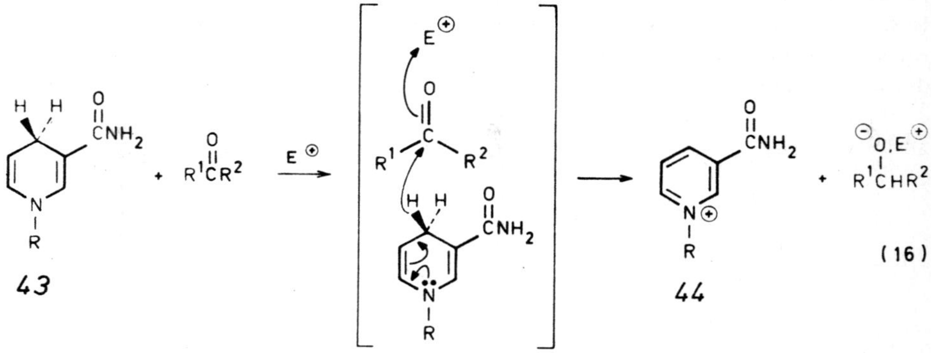

(16)

In *nonenzymic* reductions with synthetic dihydropyridines magnesium ions, usually as the perchlorate, $Mg(ClO_4)_2$, are the best electrophilic catalysts so far found. [53] The $Mg^{2\oplus}$ ion may in fact do more than just polarize the carbonyl group although this remains a point of contention. Other catalysts have been used but in general with less success than $Mg^{2\oplus}$ [54].

Two basically different strategies have been followed in combining dihydropyridines with crown ethers to obtain catalytically active systems. One approach is to attach one or more derivatives of (*43*) to or in the periphery of a crown ether, which has good complexing properties for metal ions. Those metal ions should be the electrophilic catalyst for coordination and activation of a carbonyl substrate (eq. 17). The success

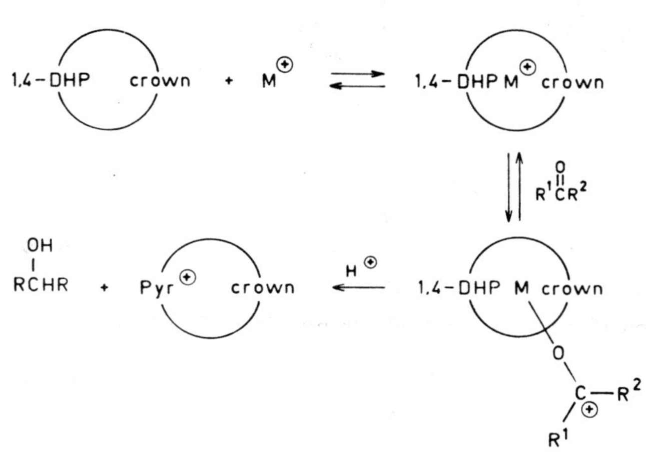

(17)

of this approach depends on a proper union of proximity effects and orientation of the dihydropyridine, electrophilic catalyst and carbonyl substrate.

A second strategy, more analogous with the syntheses of crown ethers with transacylation capabilities, is to use primary ammonium ions with the potential substrate in the substituent R of the ammonium ion; the dihydropyridine is attached to the crown ether (eq. 18). This approach poses restrictions on the types of reactions that can be carried out because there is no built-in possibility for electrophilic catalysis within the complex.

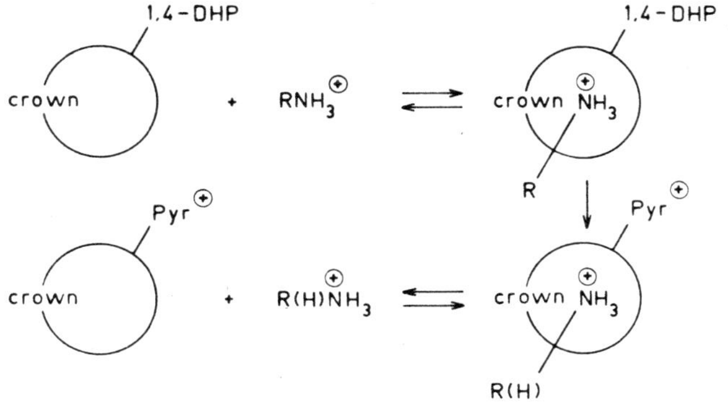

(18)

Before proceding to examples of these strategies the reader is reminded of some occasionally overlooked synthetic aspects of dihydropyridine/pyridinium salt chemistry. The pyridinium salt to 1,4-dihydropyridine conversion (44 to 43, eq. 15) is carried out virtually quantitatively and completely regioselectively with sodium dithionite, [55] and pyridinium salts are conveniently obtained from alkylation of pyridines with primary alkyl halides or similar reagents as illustrated in eq. 19. Sodium dithionite is the only reagent that carries out the pyridinium salt to 1,4-dihydropyridine reduction regioselectively. All other functionalities present in a synthetic system must be compatible with this (mild) reducing agent. [56] The dihydropyridines themselves must bear at least one electron withdrawing group

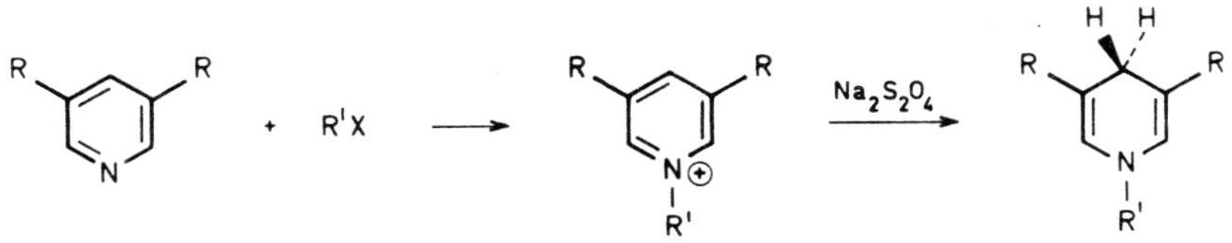

(19)

at the 3-position (for example 1,4-dihydronicotinamide 43) or two at the 3- and 5-positions (see further) to have reasonable stability. As a final point, 1,4-dihydropyridines not alkylated at nitrogen are available (for example 46, eq. 20) [57] but on hydride transfer to substrate S a pyridine remains and there are no regioselective ways of reducing a *pyridine* to a 1,4-dihydropyridine in contrast the reversibility of the pyridinium salt/N-alkyl-1,4-dihydropyridine link, which is one of its main attractions.

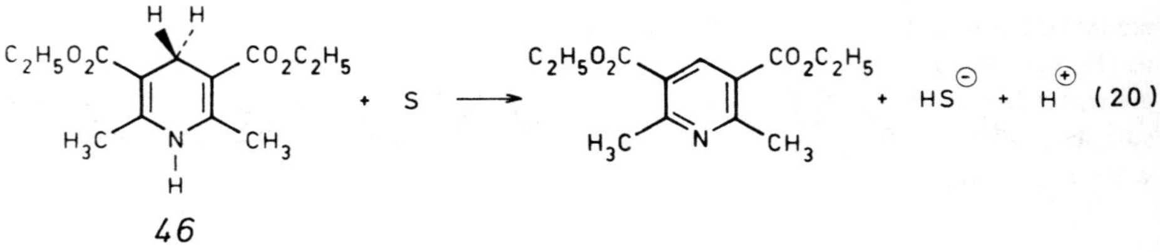

$$C_2H_5O_2C\text{-[dihydropyridine, H,H; H_3C, N-H, CH_3]}\text{-}CO_2C_2H_5 + S \longrightarrow C_2H_5O_2C\text{-[pyridinium, H_3C, N, CH_3]}\text{-}CO_2C_2H_5 + HS^- + H^+ \quad (20)$$

46

The approach of eq. 18 has been followed by Lehn's group.[58] The tartrate derived [18]crown-6 derivative (*22*) has been connected via the pyridine nitrogens to four nicotinamide derivatives to form (*47*). A potential substrate can be attached to an ammonium salt to give, for example, (*48*), which complexes in the crown ether (eq. 21). The only reaction so far investigated using this approach is, as can be seen from eq. 21, not a carbonyl group reduction but rather hydride transfer to a pyridinium salt.

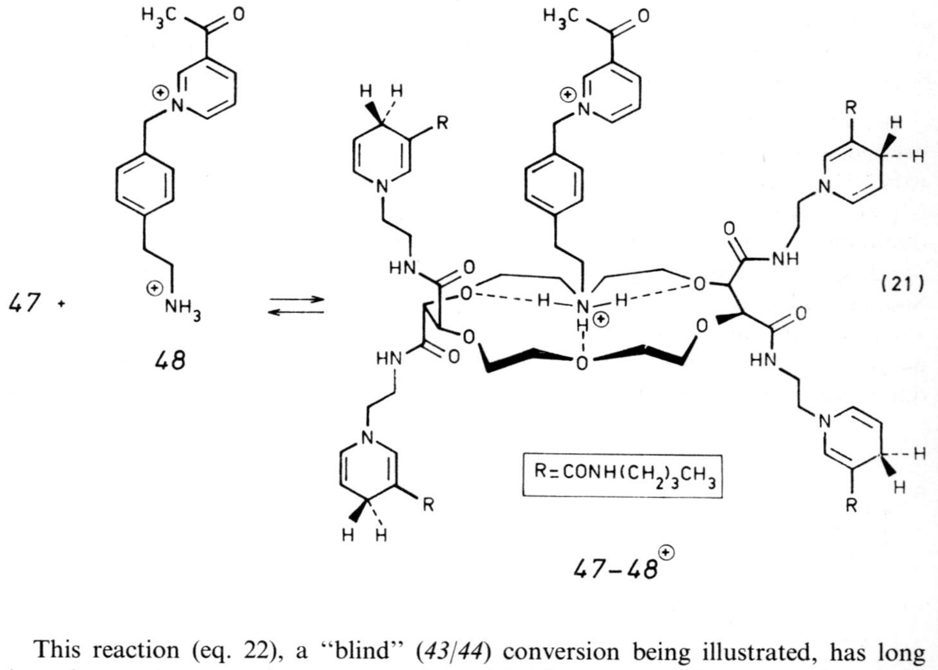

$$47 + \text{[NH}_3^+\text{]}\;\;\textbf{48} \quad \rightleftharpoons \quad 47\text{-}48^+ \qquad (21)$$

R = CONH(CH$_2$)$_3$CH$_3$

This reaction (eq. 22), a "blind" (*43/44*) conversion being illustrated, has long been known to proceed spontaneously and is also an enzymic process for transhydrogenases are known to catalyze the NADH/NAD(P)H interconversion (eq. 23).[59] The reaction involves direct transfer of "hydride" from the 1,4-dihydropyridine to the 4-position of the pyridinium salt both in the enzymic and nonenzymic examples. The position (eq. 22) of the equilibrium for 1,4-dihydropyridines and pyridinium salts with nonidentical substituents depends on the difference in reduction potentials of the two possible 1,4-dihydropyridines.[60] A considerable complication can be that the 2(6)-position of the pyridinium salt also acts as hydride accepter site, leading to isomeric dihydropyridines; this phenomenon has been studied in detail [59g,h].

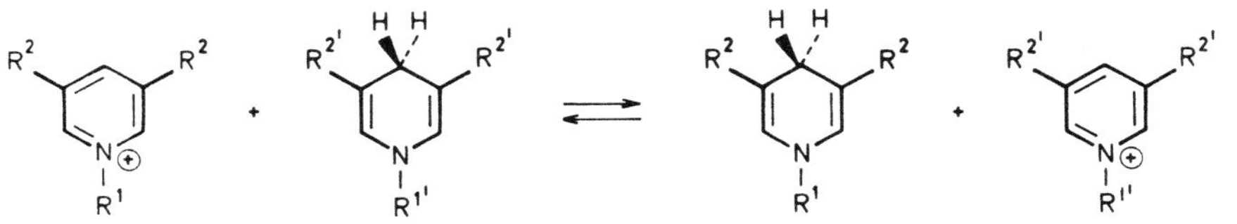

$$ \text{(22)} $$

$$ \text{NADPH} + \text{NAD}^\oplus \underset{}{\overset{\text{enzyme}}{\rightleftarrows}} \text{NADP}^\oplus + \text{NADH} \qquad (23) $$

When ammonium salt (*48*) is allowed to combine with (*47*) (eq. 21) a complex (*47–48*⊕) having presumably the indicated structure is formed. In this connection one notes also that (*48*) (as well structurally related ions) is bound to, for example, (*49*) to which tryptophane methyl ester units have been attached.[61] This leads to development of a charge-transfer absorption in the complex (*49–48*⊕). That indole-nicotinamidium interactions can provide a charge-transfer absorption is well-known [62] and has been suggested to the responsible for the color of the NAD⊕/3-phosphoglyceraldehyde dehydrogenase complex [63].

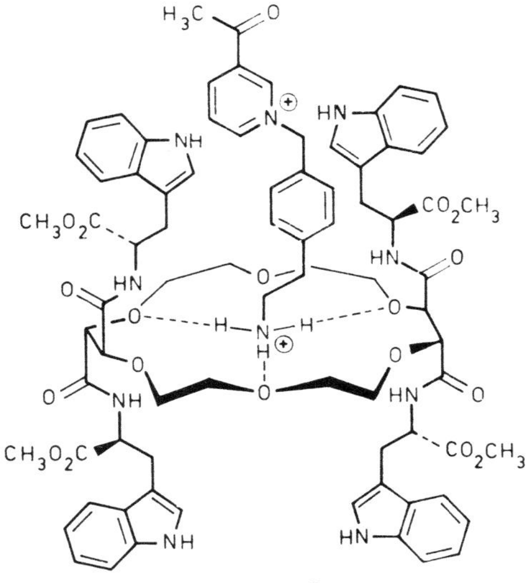

49–48⊕

Such extra bonding interactions, which might also be present in the *47–48*⊕ complex, add extra stability and provide a nice framework for studying the geometrical requirements of intermolecular interactions between groups.

The rate of hydride transfer is indeed mildly enhanced (assignment of an exact number does not seem justified) in the (47–48⊕) complex; hydride transfer is also slowed by adding K⊕ ions, which displace (48) from its binding site. A disadvantage of (47) as ligand is that the catalytically active groups probably have too much conformational freedom and do not reach an optimal geometry for hydride transfer.

Our own approach to the combination of crown ether and dihydropyridine chemistry has involved constructing the dihydropyridine as an integral portion of the macrocyclic crown ether ring (see 24b, for example). The first synthetic approach involved ring-closure of an alicyclic precursor by means of the Hantzsch 1,4-dihydropyridine synthesis as illustrated for the preparation of (50, eq. 24) [64,65]. Such "Hantzsch esters" (general type 46) are attractive in that the acid functionalities at the 3,5-positions can be used as "handles" for attaching the (dihydro)pyridine

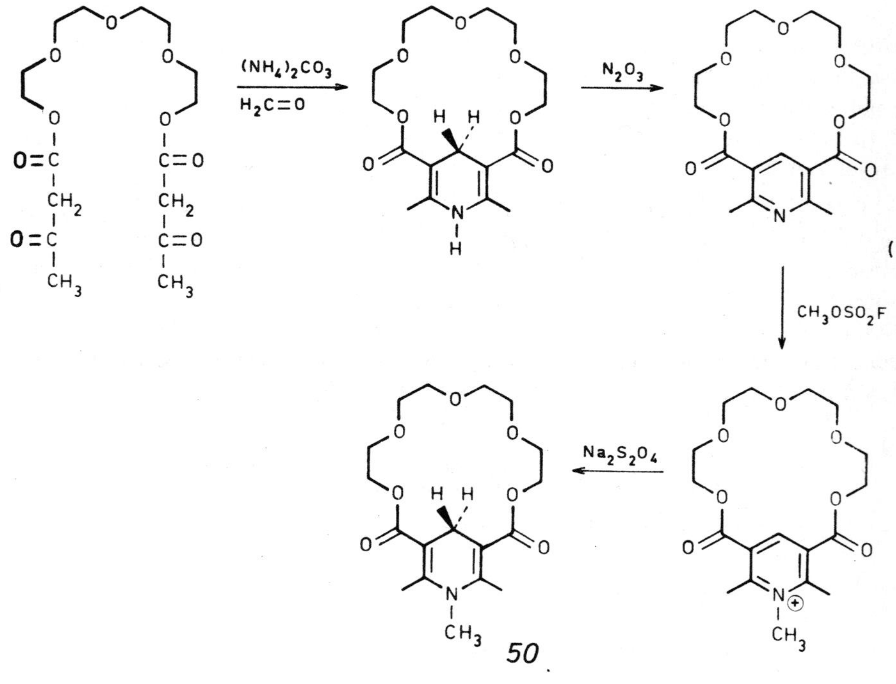

(24)

into the macrocycle ring. The success of the synthesis (20–25% yields) of eq. 24 probably lies in the use of $(NH_4)_2CO_3$, the $NH_4^⊕$ ions acting both as ammonia source [66] as well as template [24].

The 3,5-bridged 1,4-dihydropyridines like 50 are not true crown ethers owing to the drastic structural changes in the ring. The stability constants for complexation by (50) and related structures have not yet been measured but there are sufficient indications that complexing power still remains. Compound (50) complexes positive ions in solution [67] and stable complexes with $NaClO_4$ have been isolated and the structure of one complex has been determined. [68] Unfortunately (50) and related compounds in the presence of various metal ions did not give reproducible reductions of alcohols as had been hoped (eq. 18). [69] That enhanced reactivity was nevertheless present in (50) was established in an unusual reaction. Phenacyl sulfonium salts (51) can be reduced by (52) as shown in eq. 25 [70]. The reaction involves a reductive cleavage of a carbon-sulfur bond rather than reduction of the

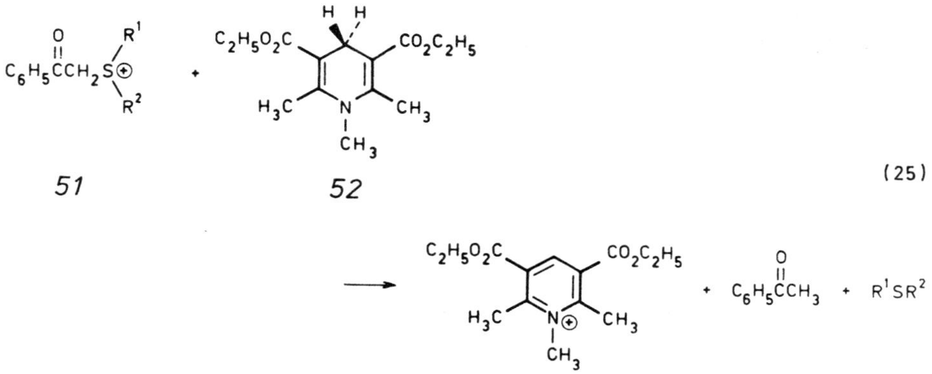

$$\underset{51}{C_6H_5\overset{O}{\overset{\|}{C}}CH_2\overset{\oplus}{S}\overset{R^1}{\underset{R^2}{\diagup}}} + \underset{52}{\underset{\underset{CH_3}{|}}{\underset{H_3C\quad N\quad CH_3}{C_2H_5O_2C\overset{H\;H}{\diagdown\diagup}CO_2C_2H_5}}} \tag{25}$$

$$\longrightarrow \underset{\underset{CH_3}{|}}{\underset{H_3C\quad \overset{\oplus}{N}\quad CH_3}{C_2H_5O_2C\diagdown\diagup CO_2C_2H_5}} + C_6H_5\overset{O}{\overset{\|}{C}}CH_3 + R^1SR^2$$

carbonyl group of (51); a chain electron transfer mechanism explains the results. [70] The bridged compound (50) carries out the same transformation of (51) ($R^1 = C_6H_5$, $R^2 = CH_3$) but, extrapolated to 75°, the reduction by (50) proceeds $2.7 \cdot 10^3$ times more rapidly than that by nonbridged (52) [69]. This accelerated reduction was ascribed to formation in a rapid preequilibrium step of a complex with the large sulfonium salt (51) perched on (50) as illustrated. Although the complexation constant with large (51) must be small, this explanation appears to be correct especially in view of the strongly inhibiting effect of $NaClO_4$ on the reduction; $Na^\oplus$ ions compete better than (51) for the binding sites in (50).

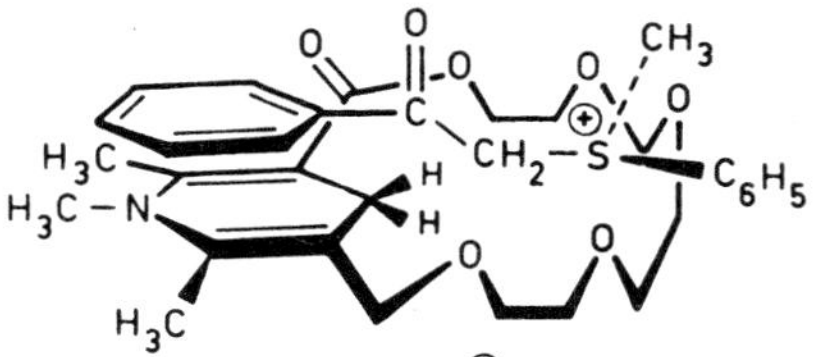

$$50\text{-}51^{\oplus}$$

Although encouraged by these results we realized that considerable redesign was in order. Moreover, difficulties in alkylating the pyridine nitrogen [65] made it necessary to remove the methyl groups at the 2,6-positions of the pyridine ring; this meant abandoning the Hantzsch approach (eq. 24).

A direction that redesign of the system could take was suggested by experiments with 1,4-dihydronicotinamide derivatives (55) with an optically active amine attached to the side chain; such compounds in the presence of $Mg^{2\oplus}$ are capable of reducing activated carbonyl compounds like ethyl phenylglyoxalate (53) to ethyl mandelate (54) with modest transfer of chirality (eq. 26) [71]. This approach has subsequently been developed by Ohno [72] into an extremely efficient method for the transfer of chirality in some reductions. We felt that the selectivity and efficiency of these derivatives of (52) could be improved by restricting the conformational flexibility by bridging as shown in (56), which with properly substituted bridges would have aspects of a crown ether. Our anticipation was that

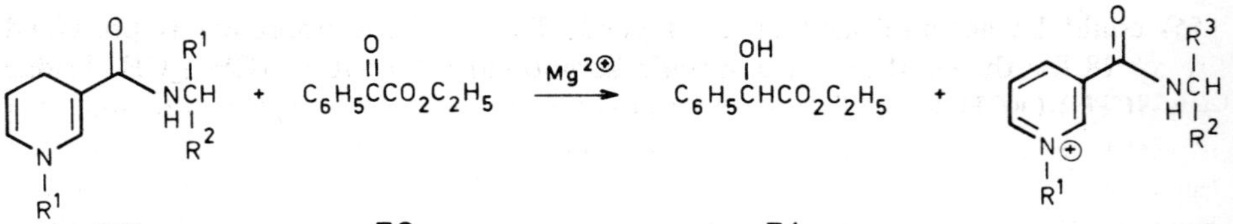

(26)

three point binding of an electrophilic catalyst would be possible using the two amide groups and a heteroatom in the bridge. If complexation of a carbonyl group occurred as illustrated in (56a) predictable enantioselectivity should result.

The chief synthetic problem in an approach to (56) would be the incorporation of the bridge thereby creating a macrocyclic ring. Numerous approaches to bridged pyridines have been described [73] but none met our needs. We therefore developed an independent route. We had been struck by the report that cesium-carboxylates in *DMF* are powerfully nucleophilic. [74] We were well rewarded on applying this report to commercially available pyridine-3,5-dicarboxylic acid (57), which could be converted into, for example, (58) in 90% yield (eq. 27) [75].

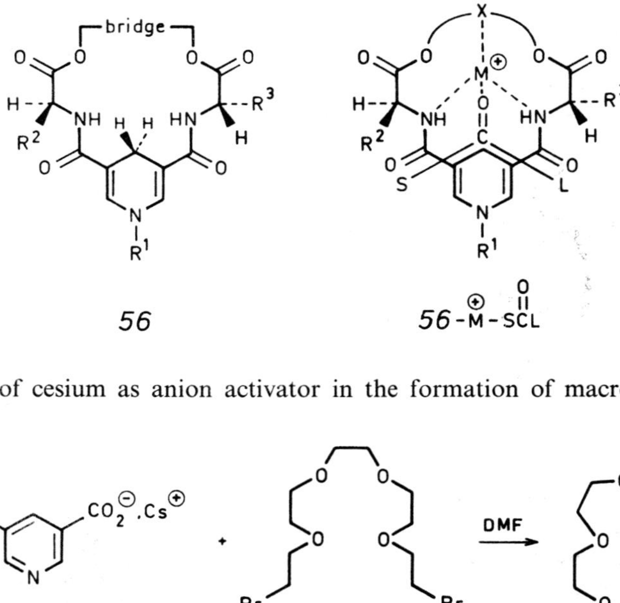

The use of cesium as anion activator in the formation of macrocycles has not

(27)

been reported previously [76]. This approach using cesium salts has subsequently been developed by us into a general method for the synthesis of macrolides, [77] macrocyclic bis-phenols, [78] macrocyclic sulfides, [79] and other applications [80].

Using the cesium salt approach the synthesis of compounds of general structure (56) could be accomplished in good yield. The synthetic approach is presented in eq. 28 for the synthesis of a specific compound (56a) [R = $(CH_3)_2CH$, bridge = $-(CH_2)_2O(CH_2)_2-$] [80]. It has been possible to prepare a series of compounds (56) in which the structural parameters of R group and length and shape of bridge have been varied. [31, 82] A considerable improvement in the synthetic scheme (eq. 28) has been the development of a direct coupling method for connecting the amino acids to the acid (eq. 29), which now allows multigram syntheses to be carried out.

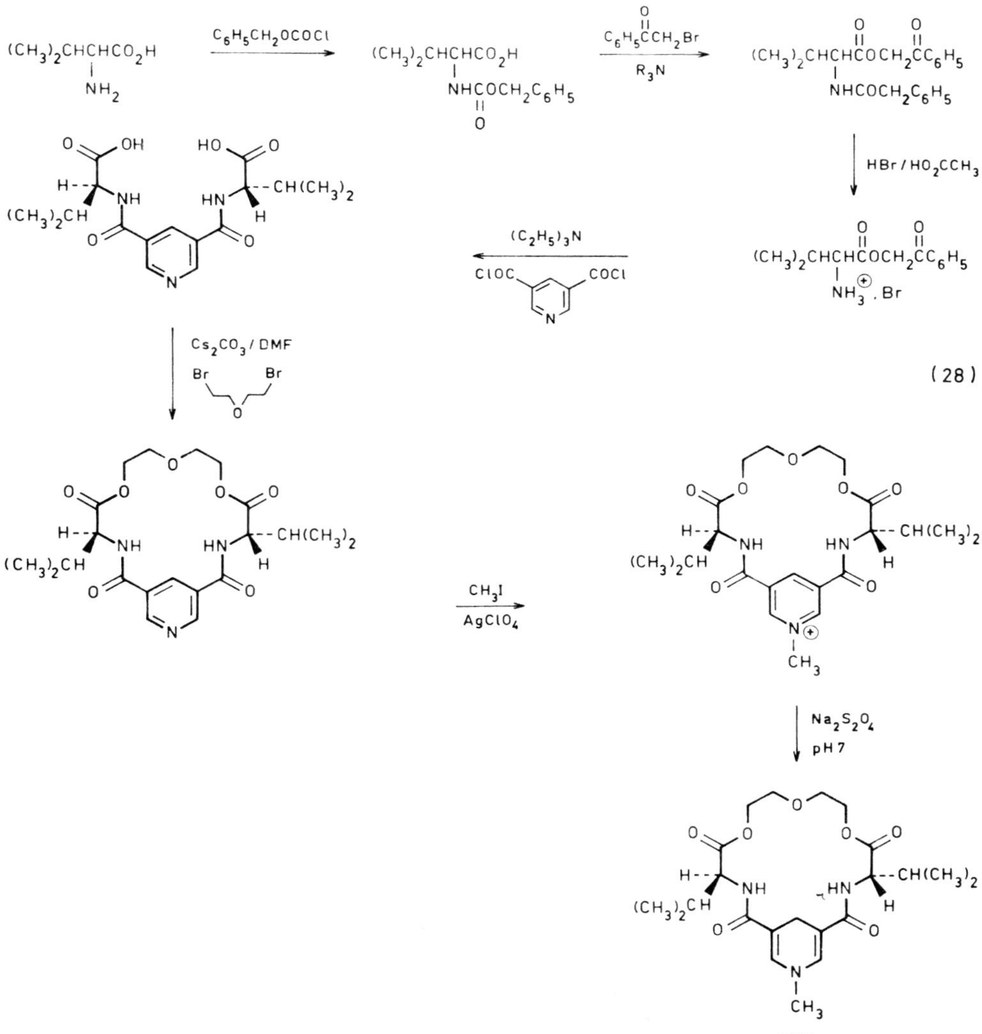

(28)

56a

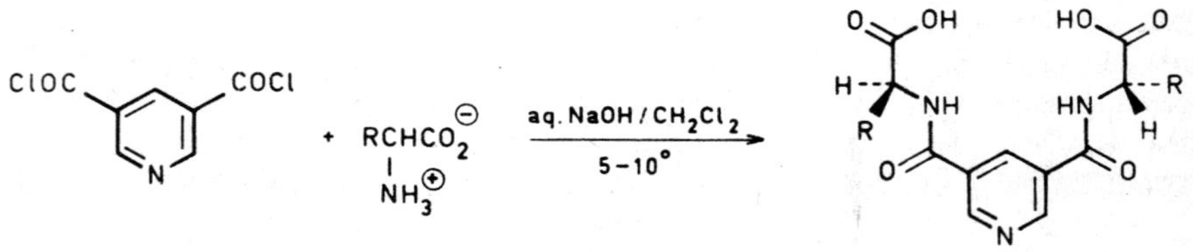

[Structures for eq. (29): pyridine-3,5-dicarbonyl chloride + RCHCO$_2^-$/NH$_3^+$ → bis-amide crown under aq. NaOH/CH$_2$Cl$_2$, 5–10°]

The compounds (56) are indeed capable of reducing activated carbonyl derivatives. The results of numerous experiments using valine, phenylalanine, and proline derived crown ethers are given in the Table [81,82]. The overall reaction is that shown in eq. 30.

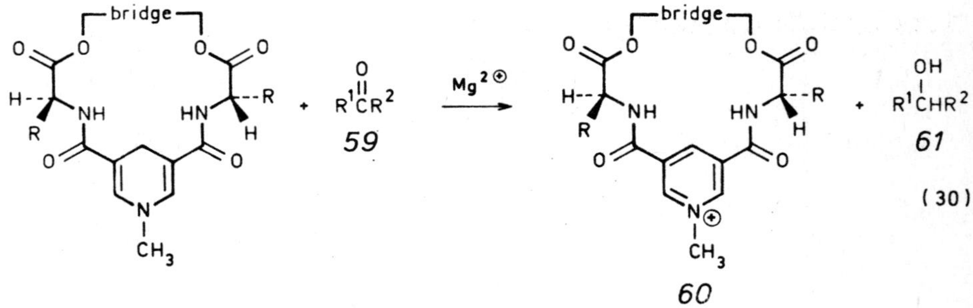

[Structures for eq. (30): dihydropyridine crown 56 + R^{1}CR^2=O (59) → pyridinium crown 60 + R^{1}CHR^{2}OH (61)]

For comparison purposes noncyclic compounds 62a, b were also examined.

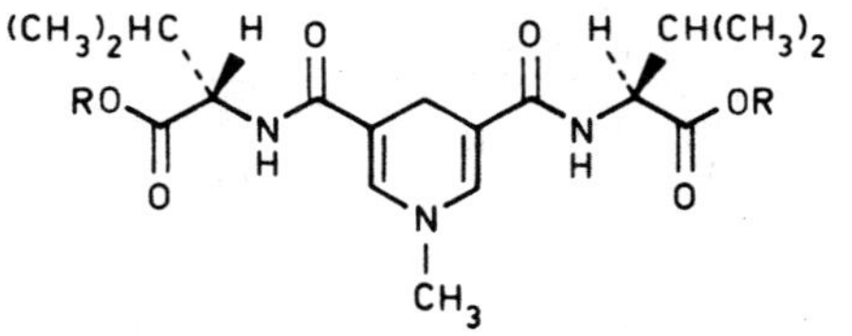

[Structure 62]

a) R = CH$_3$
b) R = CH$_3$O(CH$_2$)$_2$–

These reduce ketone (59) [R^3 = C$_6$H$_5$, R^2 = CO$_2$C$_2$H$_5$] cleanly to the corresponding alcohol (eq. 30) but in enantiomeric excesses of respectively 10% and 18% of the *R-configuration*.

The results reported in the Table are extremely promising. In contrast to the poor enantiomeric excesses obtained with (62a, b) all cyclic derivatives (56) except for proline (entry 19) give good to excellent transfer of chirality, in many cases high transfer. The results are completely stereochemically consistent in that (56) derived from L-amino-acids always produces S alcohols (the relative group priorities in all the reduced alcohols are identical allowing direct comparison). This points to similar stereochemical factors affecting the transition states for each reduction. These stereochemical factors can lead to good transfer of chirality although they

Table 1. Reduction of Activated Ketones (60) by 1,4-Dihydropyridines (56)

Entry	Amino Acid	Bridge	Enantiomeric Excess	Major Enantiomer	Substrate
1	L-valine	$-(CH_2)_2O(CH_2)-$	86%	S	$R^1=C_6H_5; R^2=CO_2C_2H_5$
2	L-valine	$-(CH_2)_2[O(CH_2)_2]_2-$	43	S	$R^1=C_6H_5; R^2=CO_2C_2H_5$
3	L-valine	$-(CH_2)_2[O(CH_2)_2]_3-$	54	S	$R^1=C_6H_5; R^2=CO_2C_2H_5$
4	L-valine	$-(CH_2)_4-$	±65	S	$R^1=C_6H_5; R^2=CO_2C_2H_5$
5	L-valine	$-(CH_2)_5-$	90	S	$R^1=C_6H_5; R^2=CO_2C_2H_5$
6	L-valine	$-(CH_2)_6-$	88	S	$R^1=C_6H_5; R^2=CO_2C_2H_5$
7	L-valine	$-(CH_2)_8-$	83	S	$R^1=C_6H_5; R^2=CO_2C_2H_5$
8	L-valine	$-(CH_2)_{10}-$	53	S	$R^1=C_6H_5; R^2=CO_2C_2H_5$
9	L-valine	$-(CH_2)_{12}-$	42	S	$R^1=C_6H_5; R^2=CO_2C_2H_5$
10	L-valine	$m-CH_2C_6H_4CH_2-$	86	S	$R^1=C_6H_5; R^2=CO_2C_2H_5$
11	D-valine	$-(CH_2)_6-$	85	R	$R^1=C_6H_5; R^2=CO_2C_2H_5$
12	L-valine	$-(CH_2)_2O(CH_2)_2-$	68	S	$R^1=C_6H_5; R^2=CF_3$
13	L-valine	$-(CH_2)_2O(CH_2)_2-$	64	S	$R^1=C_6H_5; R^2=CONH_2$
14	L-valine	$-(CH_2)_2O(CH_2)_2-$	78	S	$R^1=C_6H_5; R^2=CONHC_2H_5$
15	L-phenylalanine	$-(CH_2)_2O(CH_2)_2-$	87	S	$R^1=C_6H_5; R^2=CO_2C_2H_5$
16	L-phenylalanine	$-(CH_2)_2O(CH_2)_2-$	84	S	$R^1=C_6H_5; R^2=CONHC_2H_5$
17	L-phenylalanine	$-(CH_2)_2O(CH_2)_2-$	60	S	$R^1=m-C_6H_4OC_6H_5; R^2=CO_2CH_3$
18	L-phenylalanine	$-(CH_2)_2O(CH_2)_2-$	20	S	$R^1=m-C_6H_4OC_6H_5; R^2=CONH_2$
19	L-proline	$-(CH_2)_2O(CH_2)_2-$	0	—	$R^1=C_6H_5; R^2=CO_2C_2H_5$

do not produce in any of these systems significant rate accelerations (quantitative measurements have, however, not been carried out). All of the reductions are quite clean and give the alcohols (61) in usually about 60–80% yield after work-up and isolation (yields determined by ^{1}H-NMR are usually 85–90%).

Two of the most important conclusions to be drawn from these results are: (a) the original model (56a) used for the design of (56) must be incorrect. Comparison of entries 1–3 with entries 4–10 reveals that the enantiomeric excesses are not significantly influenced by the presence of a heteroatom as coordinating site in the bridge. (b) For bridges not longer than eight atoms (entries 1, 4–7, 10, 12–17) almost without exception irregardless of shape (entry 10) or content of the bridge high enantiomeric excesses are found.

Apparently the compounds (56) deviate so strongly from classical crown ether structures that other complexation factors — leading nevertheless to high enantioselectivity — become important. Because only weak complexes are formed in solution interpretation of the results remains speculative. We ascribe, however, the formation of S-alcohols from L-amino acid derivative (56) to the formation of a ternary complex having the structure crudely represented in (63). The main stereochemical

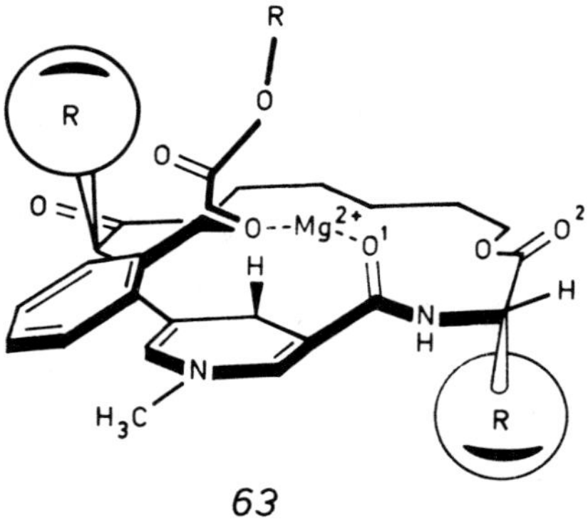

63

features are the amino acid substituents (no significant differences are seen between the isopropyl of valine compared to the benzyl of phenylalanine, both of which provide steric hindrance on *one* side of the molecule). From ^{13}C studies of the complexation of (56) with $Mg^{2\oplus}$ we conclude that 0–1 binds with the $Mg^{2\oplus}$ ion.[83] It is sterically attractive to allow 0–2 also to coordinate to $Mg^{2\oplus}$ although there is no direct spectroscopic evidence for this. This complexation of $Mg^{2\oplus}$ is *sterically directed*, i.e. it occurs specifically from the *least hindered* side of 0–1 and introduces a conformational change as the carbonyl groups move out of plane to optimal coordinating positions. The positioning of the substrate with the carbonyl carbon over the dihydropyridine "hydride" to be transferred is dictated by the steric bulk of the amino acid substituent. Examination of CPK models reveals that the phenyl group of, say, ethyl phenylglyoxalate (53) fits well only when it is placed over the dihydropyridine ring. This leads to transfer of hydride to the *re*-face of the carbonyl group producing the S-enantiomer of the alcohol in accord with experimental observation.

The drop in enantiomeric excesses of alcohol that occurs when the length of the bridge exceeds eight atoms is probably a consequence of increased conformational mobility of the macrocyclic ring, which allows the amino acid substituent to adopt conformations with it lying in the plane of the dihydropyridine ring. These conformational movements change the morphology of (56) with a consequent loss of steric discrimination for complexing of the carbonyl substrate. A similar effect is probably also involved for the proline derivative (entry 19, Table), which, owing to the 5-membered ring, adopts a heavily flattened conformation.

5.5 Macrocycles Capable of Catalyzing Other Bond-Forming or Bond-Breaking Reactions

Despite the obvious potential relatively few examples have been described of the catalysis of other reactions by macrocyclic compounds bearing reactive functional groups. One example is the promotion of macrolide formation (i.e. 65→66) [84] by functionalized crown ether (64) as shown in eq. 31. [85] Reasonable yields of macrolides are obtained for n of the order of 10 and m = 2 in (64). The intention of the approach of eq. 31 is to hold the nucleophilic alkoxide close to the carbonyl group to be attacked by chelation to $K^{\oplus}$ in the crown ether ring. Overcoming the entropy effect involved in *intramolecular* cyclization is the major problem of macrolide synthesis. [86] To accomplish this with (64) it is necessary to choose m and n so that optimal coordination can be achieved for the (presumed) tetrahedral

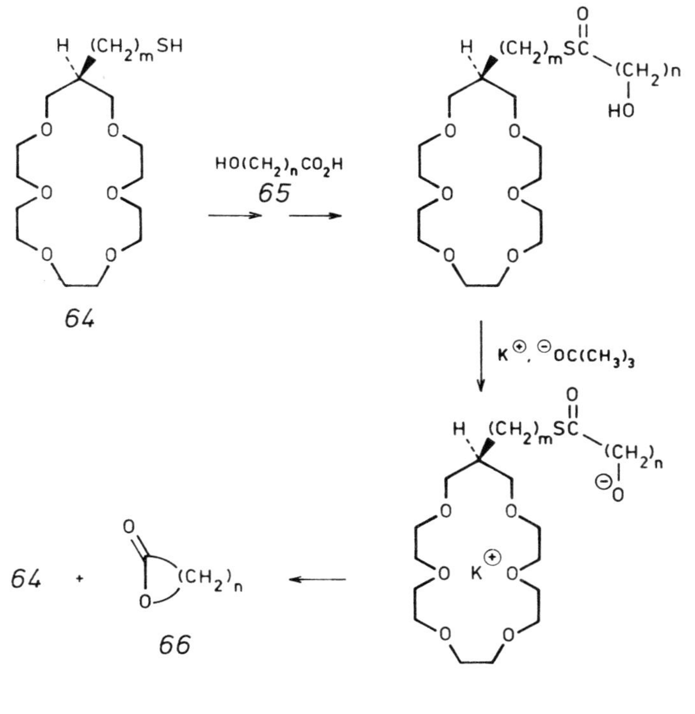

Richard M. Kellogg

intermediate formed on attack of carbonyl by alkoxide as depicted schematically in eq. 32. This amounts to complementarity in the transition state for the reaction.

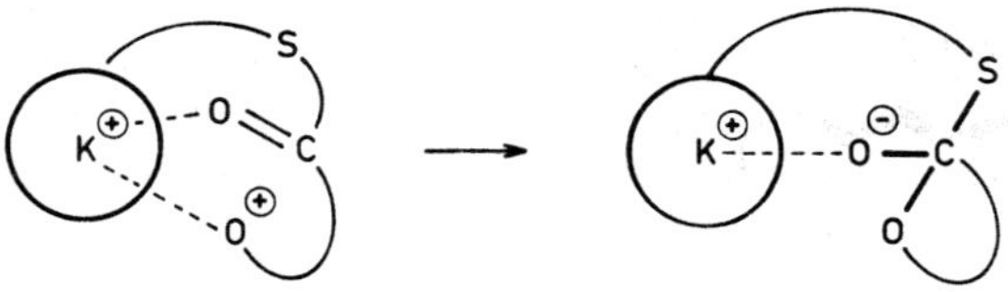

(32)

The potential for the catalysis by bioorganic modelling with macrocycles is enormous and is limited chiefly by ingenuity. For example, interesting possibilities for the future are held in sugar derived crown ethers like (23). [30] Likewise the light driven conformational changes seen in crown ethers (67) or cryptates (68) as shown in eqs. 33 and 34 offer an exciting basis for the development of catalytic systems [87,88].

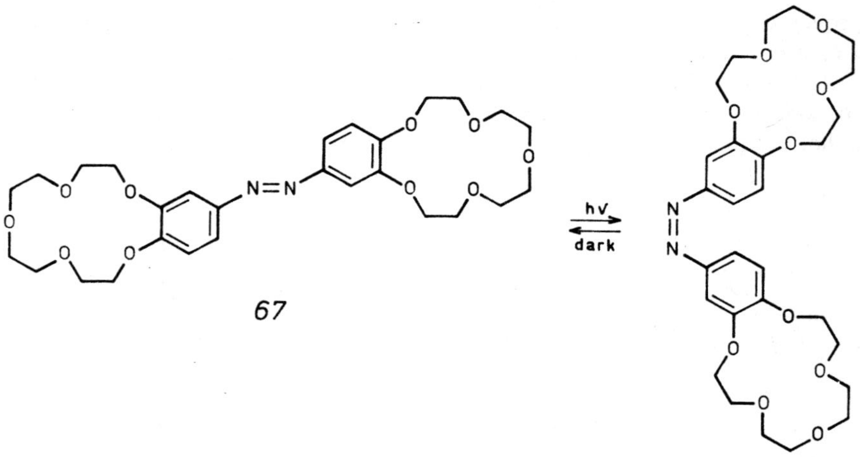

(33)

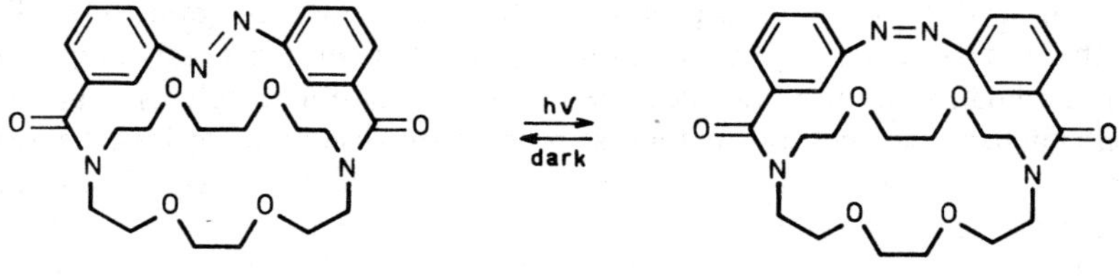

(34)

140

Other prospects are opened by binuclear complexes like (69), which is capable of complexing two copper ions. [89] The cavity between the two copper ions should

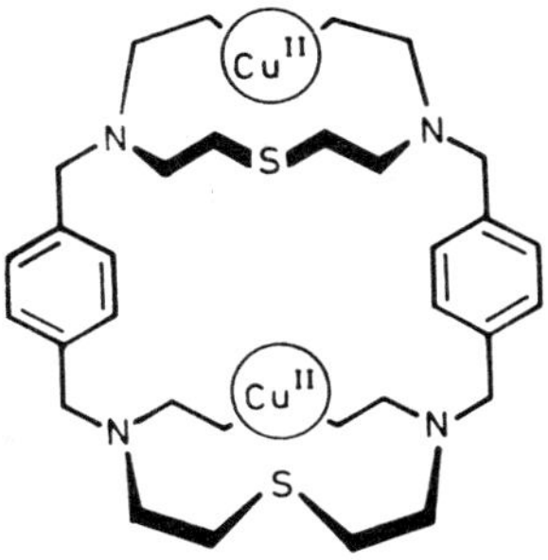

69

be large enough to accept a substrate offering thereby the opportunity to model oxygenase activity [90].

Finally, positively charged complexes linke (70) [91] are mentioned for their possibilities in the complexing and activation of anions. The development of general

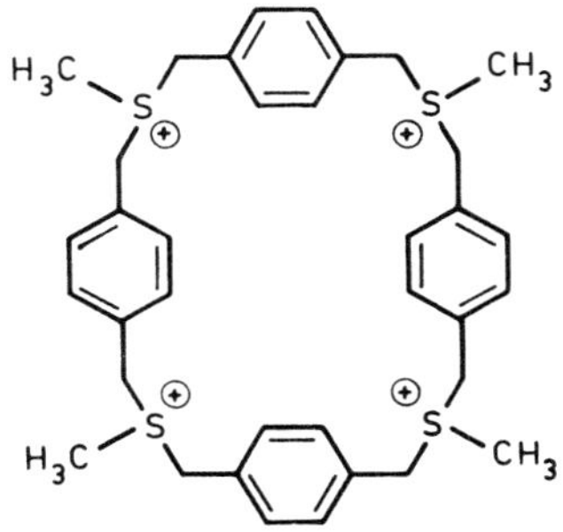

70

synthetic routes to macrocyclic sulfides with conformationally flexible chains [92] offers now the opportunity to develop new anion activating systems.

6 Summary

In terms of the reporting of accomplished chemistry this review can do no more than give an indication of the rapid progress in the branch of bioorganic modelling based on the use of macrocyclic compounds that (usually) act as complexing agents. What remains to be done, however, is to point out problems that have not been satisfactorily solved and to suggest other profitable areas of investigation.

From the material accumulated in this review one can draw the conclusion that especially crown (or cryptate) systems offer special advantages in bioorganic modelling because such compounds can — enzyme like — complex a potential substrate. On the basis of quite simple binding considerations, coupled with an analysis of steric interactions, accurate predictions of the stereochemistry of the complex can be made. The inclusion of catalytic groups in the crown (or cryptate) system and reactive functional groups in the substrate is then done in such a fashion that the stereoelectronic arrangement is compatible with the predicted geometry of the complex. However, the good complexing ability of the ligand is paradoxically often its greatest failing in terms of developing a system in which the functionalized ligand acts truly as a catalyst. As seen from much of the chemistry discussed in this review the ligand is incapable of the double task of complexing substrate but *releasing* product in an enzymic fashion, i.e. that *turnover* occurs.

How is this problem to be solved? Induced conformational changes are an obvious approach although the design of proper systems remains a challenge for which few suggestions outside of unlimited ingenuity can be given. Much of the solution to such problems will lie also in a much better understanding than we now have of *non-covalent* interactions and the stereochemistry of such interactions.[93] The assembly and disassembly of large molecular aggregates by the making and dissolution of non-covalent bonds is an art at which chemists are still relative amateurs.

A better understanding of non-covalent interactions may also provide the key to achieving also the twin goals of *both* speed and selectivity in bioorganic modelling. As far as enantioselectivity is concerned it is clear that this can be achieved fairly effectively by the use of relatively small, but appropriately placed, groups that force the substrate to complex in an enantioselective step with the ligand. In other words, the problem of enantioselectivity can be solved at the stage of *complex forming*, which is kinetically rapid. The problem of rate enhancement lies in the complex itself and can be solved effectively only by achieving optimal complementarity with the transition state of the reaction being catalyzed. Again the achievement of this goal lies in ingenuity of design.

Potential areas of applications of chiral crown ether (or cryptate) ligand systems in bioorganic modelling lie in, for example, the formation of carbon-carbon bonds, development of oxidative processes (i.e. $NAD^{\oplus}$-like mediated dehydrogenations of alcohols), synthesis of metal (perhaps binuclear) complexes with oxygenase activity, and the development of compounds with group transfer abilities. This short list includes only a few of the exciting challenges for the future that with ingenuity and imagination can be solved.

7 Acknowledgement

The work from my group has been carried out chiefly by T. J. van Bergen, J. Buter, P. Jouin, B. J. van Keulen, W. H. Kruizinga, P. Piepers, C. B. Troostwijk, and J. G. de Vries. To these coworkers and others more indirectly connected with the research reported here I owe a debt of gratitude for their enthusiasm, hard work, ideas, and good fellowship.

8 References

1. Ruzicka, L.: Über den Bau der Organischen Materie, publisher I. van Druten, Utrecht, 1926. I thank Prof. V. Prelog, E. T. H. Zürich, for making me aware of this inaugural address and Dr. A. Fürst, Hoffmann-La Roche Basel, for providing a copy

2. Lehninger, A. L.: Biochemistry, 2nd ed., Worth Publishers, Inc., New York, N.Y., (1975)
3. Pauling, L.: American Scientist 36, 58 (1948)
4. Lienhard, G. E.: Science 180, 149 (1973)
5. For example: (a) van Tamelen, E. E.: Acc. Chem. Res. 5, 152 (1975), (b) Johnson, W. S.: ibid. 1, 1 (1968)
6. (a) Robinson, R.: J. Chem. Soc. 111, 762 (1917), (b) Robinson, R.: J. Chem. Soc. 1953, 999
7. (a) Visser, C. M., Kellogg, R. M.: J. Mol. Evol. 11, 163 (1978), (b) Visser, C. M., Kellogg, R. M.: ibid. 11, 171 (1978), (c) Visser, C. M.: Naturwissenschaften 67, 549 (1980), (d) Visser, C. M.: Bioorgan. Chem. 9, 261 (1980)
8. Visser, C. M., Kellogg, R. M.: Bioorg. Chem. 6, 79 (1977)
9. Enders, D., Eichenauer, H.: Angew. Chem., Int. Ed., Engl. 18, 397 (1979)
10. (a) Meyers, A. I.: Acc. Chem. Res. 11, 375 (1978), (b) Meyers, A. I., Knaus, G., Kamata, K., Ford, M. E.: J. Am. Chem. Soc. 98, 567 (1976)
11. Hiemstra, H., Wynberg, H.: ibid. 103, 417 (1981)
12. Katsuki, T., Sharpless, K. B.: ibid. 102, 5974 (1980)
13. (a) Cramer, F., Saenger, W., Spatz, H.: ibid. 89, 14 (1967), (b) James, W. J., French, D., Rundle, R. E.: Acta Crystal. 12, 385 (1959)
14. For early work, see: (a) Cramer, F.: Angew. Chem. 64, 136 (1952), (b) Cramer, F., Dietsche, W.: Chem. Ber. 92, 1739 (1959), (c) Flohr, K., Paton, R. M., Kaiser, E. T.: J. Chem. Soc. Chem. Commun. 1971, 1621
15. (a) Bender, M. L., Komiyama, M.: Cyclodextrin Chemistry, in Reactivity and Structure in Organic Chemistry, vol. 6, New York, 1978, (b) Breslow, R.: Acc. Chem. Res. 13, 170 (1980)
16. See, for example: Jencks, W. P.: Catalysis in Chemistry and Enzymology, McGraw Hill, New Yok, N.Y. (1969)
17. For example: Storm, D. R., Strominger, J. L.: J. Biol. Chem. 248, 3940 (1973)
18. Elmore, D. T., Smith, J. J.: Biochem. J. 94, 563 (1965)
19. (a) Sheehan, J. C., Bennett, G. B., Schneider, J. A.: J. Am. Chem. Soc. 88, 3455 (1966), (b) Cruickshank, P. A., Sheehan, J. C.: ibid. 86, 2070 (1964)
20. Julia, S., Masana, J., Vega, J. C.: Angew. Chem. 92, 968 (1980)
21. Sugimoto, T., Matsumura, Y., Tanimoto, S., Okano, M.: J. Chem. Soc. Chem. Commun. 1978, 926
22. Sugimoto, T., Kokubo, T., Miyazaki, J., Tanimoto, S., Okano, M.: J. Chem. Soc., Chem. Commun. 1979, 402
23. For example: (a) Cram, D. J., Cram, J. M.: Science 183, 803 (1974), (b) Cram, D. J., Cram, J. M.: Acc. Chem. Res. 11, 8 (1978)
24. For example: (a) Busch, D. N.: Rec. Chem. Prog. 25, 107 (1964), (b) Eschenmosher, A.: Pure Appl. Chem. 20, 1 (1969)
25. Nagano, O., Kobayashi, A., Sasaki, Y.: Bull. Chem. Soc. Jpn. 51, 790 (1978)
26. Chao, Y., Weisman, G. R., Sogah, G. D. Y., Cram, D. J.: J. Am. Chem. Soc. 101, 4948 (1979) and many previous papers cited therein
27. Cram, D. J. et al.: J. Org. Chem. 43, 1930 (1978)
28. (a) Girodeau, J. M., Lehn, J. M., Sauvage, J. P.: Angew. Chem. 87, 813 (1975), (b) Behr, J. P., Lehn, J. M., Vierling, P.: J. Chem. Soc., Chem. Commun. 1976, 621
29. See, for example: Seebach, D., Hungerbühler, E. in: Modern Synthetic Methods 1980 (ed. Scheffold, R.), Otto Selle Verlag, Frankfurt, 1980, p. 91
30. Review, Stoddart, J. F.: Chem. Soc. Rev. 8, 85 (1979)
31. Jouin, P., Troostwijk, C. B., Kellogg, R. M.: J. Am. Chem. Soc. 103, 2091 (1981)
32. (a) Drenth, J. et al.: Nature 218, 929, (b) op. cit., Phil. Trans. Roy. Soc. Ser. B. 2579, 231 (1970)
33. See, for example: Walsh, C.: Enzymatic Reaction Mechanisms, W. H. Freeman and Co., San Francisco, 1979
34. (a) Behr, J. P., Lehn, J. M., Vierling, P.: J. Chem. Soc. Chem. Commun. 1976, 621, (b) Lehn, J. M.: Pure and Appl. Chem. 50, 871 (1978), (c) Behr, J. P. et al.: Helv. chim. Acta 63, 2096 (1980)
35. Matsui, T., Koga, K.: Tetrahedron Lett. 1978, 1115
36. Sasaki, S., Koga, K.: Heterocycles 12, 1305 (1979)
37. Van Keulen, B. J., Kellogg, R. M.: unpublished observations

38. Ben-Naim, A.: Hydrophobic Interactions, Plenum Press, New York, N.Y., 1980
39. Murakami, Y., Sunamoto, J., Kano, K.: Chem. Lett. *1973*, 223
40. Guthrie, J. P.: J. Chem. Soc., Chem. Commun. *1972*, 897
41. See, for example: (a) Blyth, C. A., Knowles, J. R.: J. Am. Chem. Soc. *93*, 3017, 3021 (1971), (b) Hershfield, R., Bender, M. L.: ibid. *94*, 1376 (1972)
42. Murikami, Y. et al.: Bull. Chem. Soc. Japan *50*, 3365 (1977)
43. Murakami, Y. et al.: J. Chem. Soc., Perkin I *1979*, 1560
44. White, H. B., III: Pyridine Nucleotide Coenzymes (ed. J. Everse, B. M. Anderson, K. S. You) in press
45. For example: Fischli, A., Süss, D.: Helv. chim. Acta *62*, 48 (1979)
46. Fischli, A.: ibid. *62*, 882 (1979)
47. Fischli, A., Süss, D.: ibid. *62*, 2361 (1979)
48. Fischli, A., Müller, P. M.: ibid. *63*, 529 (1980)
49. (a) Fischli, A., Müller, P. M.: ibid. *63*, 1619 (1980), (b) Fischli, A., Daly, J. J.: ibid. *63*, 1628 (1980) and references cited therein
50. For example: (a) Karrer, P. et al.: Helv. chim. Acta *20*, 55 (1937), (b) Rafter, G. W., Colowich, S. P.: J. Biol. Chem. *209*, 773 (1954), (c) Abeles, R. H., Hutton, R. F., Westheimer, F. H.: J. Am. Chem. Soc. *79*, 712 (1957), (d) Mauzerall, D., Westheimer, F. H.: ibid. *77*, 2261 (1955), (e) review of dihydropyridines: Eisner, U., Kuthan, J.: Chem. Rev. *72*, 1 (1972)
51. (a) Jones, J. B., Taylor, K. E.: Can. J. Chem. *54*, 2974, 2969 (1976), (b) Taylor, K. E., Jones, J. B.: J. Am. Chem. Soc. *98*, 5687 (1976)
52. For discussions of enzyme action, see for example: (a) Eklund, H. et al.: FEBS Lett. *44*, 200 (1974), (b) Adams, M. J. et al.: Proc. Natl. Acad. Sci. USA *1974*, 1968, (c) Hill, E. et al.: J. Mol. Biol. *72*, 577 (1972), (d) Lazdunski, M.: in Prog. Bioogr. Chem. *3*, 81 (1974)
53. For example: (a) Ohno, A. et al.: J. Chem. Soc., Chem. Commun. *1978*, 328, (b) Gase, R. A., Pandit, U. K.: J. Am. Chem. Soc. *101*, 1979 (1979)
54. For example: Baba, N., Oda, J., Inouye, Y.: J. Chem. Soc., Chem. Commun. *1980*, 815 and references cited therein
55. (a) Yarmolinsky, M. B., Colowick, S. P.: Biochim. Biophys. Acta *20*, 177 (1956), (b) Biellmann, J. F., Callot, H. J.: Bull. Soc. Chim. Fr. *1968*, 1154
56. de Vries, J. G., Kellogg, R. M.: J. Org. Chem. *45*, 4126 (1980)
57. For some applications, see: Pandit, U. K. et al.: J. Chem. Soc., Chem. Commun. *1974*, 627 and references cited therein
58. Behr, J. P., Lehn, J. M.: J. Chem. Soc., Chem. Commun. *1978*, 143
59. For example: (a) Colowick, S. P. et al.: J. Biol. Chem. *195*, 95 (1952), (b) Kaplan, N. O., Colowick, S. P., Neufeld, E. F.: ibid. *195*, 107 (1952), (c) Kaplan, N. O., Colowick, S. P., Neufeld, E. F. ibid., *205*, 1 (1953), (d) Kaplan, N. O., Colowick, S. P., Neufeld, E. F.: Fed. Proc. Fed. Am. Soc. Exp. Biol. *11*, 238 (1952), (e) Cilento, G.: Arch. Biochem. Biophys. *88*, 352 (1960), (f) Spiegel, M. J., Drysdale, G. R.: J. Biol. Chem. *235*, 2498 (1960), (g) Ludowieg, J., Levy, A.: Biochemistry *3*, 373 (1964), (h) van Bergen, T. J., Mulder, T., Kellogg, R. M.: J. Am. Chem. Soc. *98*, 1960 (1976), (i) Bergen, T. J., Mulder, T., van der Veen, R. A., Kellogg, R. M.: Tetrahedron *34*, 2377 (1978)
60. These hydride exchange processes can also be used to establish redox potentials for 1,4-dihydropyridines: Piepers, O., Kellogg, R. M.: publication in preparation
61. Behr, J. P., Lehn, J. M.: Helv. chim. Acta *63*, 2112 (1980)
62. Ash, R. P., Herriott, J. R., Deranleau, D. A.: J. Am. Chem. Soc. *99*, 4471 (1977)
63. Kosower, E. M.: ibid. *78*, 3497 (1956). But see also: Piepers, O., Kellogg, R. M.: J. Chem. Soc., Chem. Commun. *1980*, 1154
64. Hantzsch, A.: Liebigs Ann. Chem. *215*, 1 (1882)
65. Kellogg, R. M. et al.: J. Org. Chem. *45*, 2854 (1980)
66. Haley, C. A. C., Maitland, P.: J. Chem. Soc. *1951*, 3155
67. van Bergen, T. J., Kellogg, R. M.: J. Am. Chem. Soc. *99*, 3882 (1977)
68. van der Veen, R. H., Kellogg, R. M., Vos, A., van Bergen, T. J.: J. Chem. Soc., Chem. Commun. *1978*, 923
69. van Bergen, T. J., Kellogg, R. M.: unpublished
70. van Bergen, T. J., Hedstrand, D. M., Kruizinga, W. H., Kellogg, R. M.: J. Org. Chem. *44*, 4953 (1979)

71. (a) Pandit, U. K., Mas Cabré, F. R.: J. Chem. Soc., Chem. Commun. *1971*, 552, (b) Ohnishi, Y., Kagami, M., Ohno, A.: J. Am. Chem. Soc. *97*, 4766 (1975), (c) Endo, T., Hayashi, Y., Okawara, M.: Chem. Lett. *1977*, 391, (d) Endo, T., Kawasaki, H., Okawara, M.: Tetrahedron Lett. *1979*, 23, (e) Ohnishi, Y., Numakunai, T., Ohno, A.: ibid. *1975*, 3813, (f) van Ramesdonk, H. J., Verhoeven, J. W., Pandit, U. K., de Boer, T. J.: Rec. trav. Chim. Pays-Bas *97*, 195 (1978)
72. (a) Ohno, A., Ikeguchi, M., Kimura, T., Oka, S.: J. Chem. Soc., Chem. Commun. *1978*, 328, (b) Ohno, A., Ikeguchi, M., Kimura, T., Oka, S.: J. Am. Chem. Soc. *101*, 7036 (1979)
73. For examples and literature citations, see: (a) Newkome, G. R., Kawato, T.: *44*, 2693 (1979), (b) Newkome, G. R., Kawato, T., Nayak, A.: ibid. *44*, 2697 (1979), (c) Heimann, U., Vögtle, F.: Angew. Chem. *90*, 211 (1978), (d) Izatt, R. M. et al.: J. Am. Chem. Soc. *99*, 2365 (1977), (e) Newkome, G. R. et al.: Chem. Rev. *77*, 513 (1977), (f) Bradshaw, J. S., Stott, P. E.: Tetrahedron *36*, 461 (1980)
74. Wang, S. S. et al.: J. Org. Chem. *42*, 1286 (1977)
75. Piepers, O., Kellogg, R. M.: J. Chem. Soc., Chem. Commun. *1978*, 383
76. But see the use of CsF in the synthesis of crown ethers: Reinhoudt, D. N., de Jong, F., Tomassen, H. P. M.: Tetrahedron Lett. *1979*, 2067
77. Kruizinga, W. H., Kellogg, R. M.: J. Chem. Soc., Chem. Commun. *1979*, 286, J. Am. Chem. Soc. in press
78. van Keulen, B. J., Kellogg, R. M., Piepers, O.: J. Chem. Soc., Chem. Commun. *1979*, 235
79. Buter, J., Kellogg, R. M.: ibid. *1980*, 466
80. Strijtveen, B., Vriesma, B., Kellogg, R. M.: unpublished
81. de Vries, J. G., Kellogg, R. M.: J. Am. Chem. Soc. *101*, 2759 (1978)
82. Jouin, P., Troostwijk, C. B., Weremus Buning, G., Talma, A. G.: publication in preparation
83. Jouin, P., de Vries, J. G., Kellogg, R. M.: unpublished
84. Interest in macrolide formations is intense. For recent reviews, see: (a) Nicolau, K. C.: Tetrahedron *23*, 683 (1977), (b) Masamune, S., Bates, G. S., Corcoran, J. W.: Angew. Chem. Int. Ed. *16*, 585 (1977), (c) Back, T. G.: Tetrahedron *33*, 3041 (1977)
85. (a) Rastetter, W. H., Phillion, D. P.: Tetrahedron Lett. *1979*, 1469, (b) Rastetter, W. H., Phillion, D. P.: J. Org. Chem. *45*, 1535 (1980)
86. See, for another approach ref. 77
87. (a) Shinkai, S., Ogawa, T., Kusano, Y., Manabe, O.: Chem. Lett. *1980*, 283, (b) Shinkai, S. et al.: Tetrahedron Lett. *1979*, 4569, (c) Shinkai, S. et al.: J. Chem. Soc., Chem. Commun. *1980*, 375, (d) Shinkai, S. et al.: Tetrahedron Lett. *1980*, 4463, (e) Shinkai, S. et al.: J. Am. Chem. Soc. *102*, 5860 (1980)
88. For a somewhat related application in cyclodextrin chemistry, see: Ueno, A., Takahashi, K., Osa, T.: J. Chem. Soc., Chem. Commun. *1980*, 837. A somewhat different application is described by: Kojima, M., Toda, F., Hattori, K.: Tetrahedron Lett. *1980*, 2721
89. Kahn, O. et al.: J. Am. Chem. Soc. *102*, 5935 (1980)
90. See, for example: Vanneste, W. H., Zuberbühler, A.: in Molecular Mechanisms of Oxygen Activation (ed. O. Hayaishi), Academic Press, New York, N.Y., 1974, pp. 371–404
91. Tabushi, I., Sasaki, H., Kuroda, Y.: J. Am. Chem. Soc. *98*, 5727 (1976)
92. Buter, J., Kellogg, R. M.: J. Chem. Soc., Chem. Commun. *1980*, 466
93. See, for example, the interview with Lord Todd: Chem. Eng. News. *58*, 28, Oct. 6 (1980), and also Lehn, J. M.: Structure and Bonding *16*, 1 (1973)

Phase-Transfer Catalyzed Reactions

Fernando Montanari, Dario Landini, Franco Rolla

Centro C.N.R. and Istituto di Chimica Industriale, Universita', Via C. Golgi 19, Milano, Italy

Table of Contents

1 Introduction 149

2 General Principles and Mechanism 150
  2.1 General Principles 150
  2.2 Partition of the Anions and/or Ion Pairs Between Aqueous and
     Organic Phases 151
    2.2.1 Extraction Equilibria 151
    2.2.2 Lipophilicity of the Cation 152
    2.2.3 Influence of the Anion 152
    2.2.4 Influence of Solvents 154
    2.2.5 Salt Effect 155
  2.3 Mechanism of Phase-Transfer Catalysis 155
    2.3.1 Fundamental Aspects 155
    2.3.2 Anionic Reactivity of Quaternary Onium Salts 157
    2.3.3 Dependence of Anionic Reactivity on the Structure of the
        Onium Cation 159
    2.3.4 Mechanism in the Presence of Concentrated Aqueous Bases . . . 161

3 Ion Pair Extraction 163

4 Solid-Liquid and Solid-Gas Phase-Transfer Catalysis 163

5 Other PTC Catalysts 164
  5.1 Crown Ethers 164
  5.2 Cryptands 167
  5.3 Podands and Polypodands 170
  5.4 Other Catalytic Systems 172

6 Phase-Transfer in the Presence of Catalysts Bonded to a Polymeric Matrix . . . 173
   6.1 Introduction 173
   6.2 Quaternary Onium Salts. The Influence of Structural Factors
       and of the Spacer Chain 173
   6.3 Crown Ethers and Cryptands 176
   6.4 Cosolvents Bonded to Polymeric Matrices 177
   6.5 Swelling of the Resin and Diffusive Factors 178
   6.6 Reaction Mechanism 178
   6.7 PTC Catalysts Bonded to Silica 179
   6.8 Uses and Chemical Stability 179

7 Applications of Phase-Transfer Catalysis 180
   7.1 Introduction 180
   7.2 Displacement Reactions 181
   7.3 Alkylations 182
   7.4 Eliminations 184
   7.5 α-Eliminations: Carbenes and Analogeous Compounds 184
   7.6 Other Base-Promoted Reactions: Isomerizations, H/D Exchanges,
       Nucleophilic Additions, Condensations 186
   7.7 Reactions Promoted by Aqueous Hydrohalic Acid 188
   7.8 Oxidations 188
       7.8.1 Permanganate 188
       7.8.2 Chromate 189
       7.8.3 Hypochlorite 189
       7.8.4 Osmium and Ruthenium Tetroxides 190
       7.8.5 Hydrogen Peroxide, Peroxides and Oxygen 191
   7.9 Reductions 191
       7.9.1 Complex Hydrides 191
       7.9.2 Other Reducing Agents 192
   7.10 Applications of PTC to Transition Metal Chemistry 192

8 Asymmetric Syntheses 193

9 References 194

1 Introduction

A number of organic reactions proceed through the attack of anions on the substrate or through primary formation of reactive anionic species. To accelerate such reactions and thus allow them to be run under mild conditions has been a traditional goal in organic chemistry. The observation that nearly unsolvated anions and/or interacting sparingly with neighboring cations are highly reactive suggested a solution of this problem.

The introduction of dipolar aprotic solvents [1] and the discovery of macrocyclic (crown ethers [2,3]) and macrobicyclic polyethers (cryptands [4,5]) represent some of the more significant steps. Dating from the late sixties, a new general technique was developed: phase-transfer catalysis (PTC). PTC has the advantages of being extremely simple and economical and so met with immediate success in industrial applications.

Reactions are conducted in a two-phase system consisting of mutually insoluble aqueous and organic layers. Ionic reagents (i.e. salts, bases or acids) are dissolved in the aqueous phase, the substrate in the organic one (liquid-liquid PTC). Alternatively, ionic reagents can be used in the solid state as a suspension in the organic medium (solid-liquid PTC). The transport of the anions from the aqueous or solid phase to the organic one, in which the reaction occurs, is ensured by catalytic amounts of lipophilic transport agents, usually quaternary onium salts.

In the absence of the latter, the reactions proceed at a very low rate or not at all. In the simplest case dealing with nucleophilic substitution reactions, PTC can be schematized as follows (Eq. (1)).

$$R-X_{org} + Y^-_{aq} \xrightarrow{\text{transport agent}} R-Y_{org} + X^-_{aq} \quad (1)$$

The introduction of PTC is mainly due to the work of three independent research groups; they operated nearly contemporaneously on similar or complementary research lines: M. Makosza, of the Technical University of Warsaw, Poland [6,7], A. Brändström, of the AB Hässle, Sweden [8,9], and C. M. Starks, of the Continental Oil Company, Ponca City, USA [10-12].

The term "phase-transfer catalysis" was introduced by Starks [10-12], and explains the fundamental mechanism of these reactions.

Reactions under PTC conditions were studied previously as well [13]; an old example is in a German patent dated 1913 [14]. In no case, however, were the general character and the practical application possibilities of this technique recognized. Even when the mechanism of a single reaction was recognized [15] did this remain an isolated observation.

Starks' major contribution was the identification of the mechanism and the determination of the scope of the method [10-12]. His fundamental work created immense interest among university and industrial researchers, and within a few years hundreds of works were reported on the matter.

A number of reviews have appeared [7,9,16-34] and very recently some excellent books as well, stressing various fields. Brändström's book [9] deals with mechanistic

149

and practical aspects of "ion pair extraction", i.e. of the stoichiometric transfer and of the reactivity of ion pairs in low polarity media. Weber and Gokel treat PTC mainly from the preparative point of view [35]. Starks and Liotta [36] analyze the mechanism of PTC, with emphasis on practical applications, reporting them in full detail in many cases. Dehmlow's volume [37] is developed along similar lines and is an exhaustive review of the matter up to the end of 1977, with many additional references up to 1979. The latter three volumes also give a wide compendium of the chemistry of macrocyclic and macrobicyclic polyethers; other reviews, peculiar of the latter systems, deal with their application in the field of anionic activation and PTC [38-47].

In addition, other techniques have recently been developed, involving base activation in heterogeneous media; particularly noteworthy are the sodium amide-sodium t-butoxide-tetrahydrofuran system by Caubère [48] and the carbon tetrachloride-t-butanol-potassium hydroxide system by Meyers [49,50]. These methods, however, do not require the presence of a catalyst for ion transport from one phase to the other, and cannot be classified as reactions carried on under PTC conditions.

2 General Principles and Mechanism

2.1 General Principles

Consider the simple case of an aliphatic nucleophilic substitution carried out in an aqueous-organic two-phase system in the presence of catalytic amounts of a quaternary ammonium or phosphonium salt Q^+Y^-. The detailed mechanism of the reaction, indicated by Eq. (2) as originally proposed by Starks [11], in which Q^+ and M^+ are the organic and inorganic cations, respectively, involves various factors including:

a) partition of the quaternary salt (or of another transport agent) between the aqueous and organic phases; or, alternatively, partition of the anions between the two phases, if Q^+ is exclusively in the organic phase;
b) structure of catalyst (and eventually its aggregation state);
c) reactivity of ion pairs in low-polarity organic medium;
d) reaction kinetics;
e) hydration state of the anions in the organic phase;
f) comparison of the reactivity under PTC conditions with that under different reaction conditions.

$$\begin{array}{ll} R-X + Q^+Y^- \rightarrow R-Y + Q^+X^- & \text{organic phase} \\ \updownarrow \qquad\qquad \updownarrow & \\ M^+X^- + Q^+Y^- \rightleftharpoons M^+Y^- + Q^+X^- & \text{aqueous phase} \end{array} \qquad (2)$$

2.2 Partition of the Anions and/or Ion Pairs Between Aqueous and Organic Phases

2.2.1 Extraction Equilibria

The extraction equilibrium (3) is regulated by equilibria (4) and (5).

$$[Q^+X^-]_{org} + M^+_{aq} + Y^-_{aq} \rightleftharpoons [Q^+Y^-]_{org} + M^+_{aq} + X^-_{aq} \tag{3}$$

$$Q^+_{aq} + X^-_{aq} \rightleftharpoons [Q^+X^-]_{org} \tag{4}$$

$$Q^+_{aq} + Y^-_{aq} \rightleftharpoons [Q^+Y^-]_{org} . \tag{5}$$

The stoichiometric extraction constant E_{QX} (6) can be deduced from (4), and the stoichiometric extraction constant E_{QY} (7) [51,52] can be analogously deduced from (5). However, for more accurate measurements, the thermodynamic extraction constants should be determined, using activities instead of concentrations.

$$E_{QX} = \frac{[Q^+X^-]_{org}}{[Q^+]_{aq}[X^-]_{aq}} \tag{6}$$

$$E_{QY} = \frac{[Q^+Y^-]_{org}}{[Q^+]_{aq}[Y^-]_{aq}} \tag{7}$$

The extraction equilibria (6) and (7) should be influenced by stoichiometric concentrations of the anion and cation in the aqueous phase. In fact, however, they also depend on many other factors, including:
a) dissociation or association degree of ion pairs in the organic and aqueous phases;
b) pH of the aqueous phase which determines the actual concentration of the anion or the cation Q^+, if the latter is a protonated base HB^+ (Eqs. (8) and (9)):

$$X^- + H_3O^+ \rightleftharpoons HX + H_2O \tag{8}$$

$$HB^+ + H_2O \rightleftharpoons B + H_3O^+ \tag{9}$$

c) the possible formation of associated ions such as $Q^+HX_2^-$ in the organic phase.
A detailed discussion on the quantitative evaluation of these factors is reported by Brändström [25].

In most cases, under PTC conditions, a competitive extraction of two or more anions occurs. For example, in a nucleophilic displacement such as (2), the extraction equilibrium (3) must be taken into account. From Eqs. (6) and (7), Eq. (10) is derived. From this, if E_{QX} and E_{QY} are known, it

$$K^{sel}_{Y/X} = \frac{E_{QY}}{E_{QX}} \cdot \frac{[Q^+Y^-]_{org} \cdot [X^-]_{aq}}{[Q^+X^-]_{org} \cdot [Y^-]_{aq}} \tag{10}$$

is possible to calculate the selectivity coefficients $K^{sel}_{Y/X}$, and from these the actual concentrations of the reactive species Q^+Y^- in the organic medium, in competition with the Q^+X^- species derived from the leaving group X^-, for any known concentration of X^-, Y^- and Q^+ and at any moment of the reaction.

Generally it is possible to use concentrations disregarding activities. If Q^+ is sufficiently lipophilic to be effectively present only in the organic phase and if side factors a)–c) are also ignored, then the fractions of Q^+ associated with X^- and Y^- may be readily calculated [23]. In many cases, satisfactory predictions for reactions under PTC conditions can be obtained in this way, at least from the preparative point of view. It should however be taken into account that side processes cannot always be neglected. Brändström reported a number of examples and quantitative evaluations of this matter [25].

2.2.2 Lipophilicity of the Cation

The lipophilicity of cation Q^+ increases with rising number of carbon atoms bonded to the cationic center. Linear correlations have been found between $\log E_{QX}$ and the number of carbon atoms n of the ammonium cation, according to the general expression (11) in which constants C and A primarily

$$\log E_{QX} = C + An \tag{11}$$

depend on the nature of the cation. For example, in the case of quaternary ammonium picrates, $A = 0.54$ in some solvents [53,54].

Such expressions are obviously of value only within homologous series. Ligands such as aromatic and benzylic groups also contribute to the lipophilicity of the cation, but in a very different manner from alkyl chains.

2.2.3 Influence of the Anion

The possibility of extracting a quaternary salt from an aqueous phase into an organic one is particularly dependent on the nature of the anion. It is strictly related to various factors, such as its hydration sphere, electronegativity, volume, structure, etc. Table 1 reports the extraction constants E_{QX} in water/chloroform and water/dichloromethane for some tetrabutylammonium salts [25,55].

It is clear that the absolute values also depend on the combination of all the other factors involved (structure of the cation, organic solvent, ionic strength of the aqueous phase), but in general the scale of the relative extractive capability is nearly constant within a given series of anions and is independent of other factors. This can be indicated as follows [55,56]:

Picrate $\gg ClO_4^- > I^- > p\text{-TosO}^- > NO_3^- > Br^- > PhCO_2^-$
$> (PhO^-, Cl^-, CN^-) > HSO_4^- > HCO_3^- > CH_3CO_2^-$
$> (F^-, OH^-, SO_4^{2-}, CO_3^{2-}) > PO_4^{3-}$.

Gordon has tabulated a series of selectivity coefficients $K_{Y/Cl}^{sel}$ relative to Cl^-, measured in various solvents for several quaternary cations on the basis of potentials developed by anion-exchange membrane electrodes Q^+Y^- in the presence of foreign anions X^- [56].

Table 1. Extraction constants (E_{QX}) of tetrabutylammonium salts in chloroform-water and methylene chloride-water [25, 55]

Anion	$CHCl_3$-H_2O	CH_2Cl_2-H_2O
OH^-	—	5.0×10^{-4}
HSO_4^-	—	6.0×10^{-2}
HCO_3^-	—	6.0×10^{-3}
$H_2PO_4^-$	—	6.0×10^{-5}
Cl^-	0.8	0.35
Br^-	19.5	17
I^-	1,023	2,188
NO_3^-	24.5	79
ClO_4^-	3,020	43,700
CH_3COO^-	7.6×10^{-3}	—
$C_6H_5COO^-$	2.45	—
C_6H_5-O^-	0.93	—
2,4,6-$(NO_2)_3C_6H_2O^-$	810,000	4,800,000
p-CH_3-C_6H_4-SO_3^-	214	—

From them, using Eq. (12), it is possible to calculate selectivity coefficients relative to any anion pair.

$$\log K_{Y/X}^{sel} = \log (K_{Y/Cl}^{sel} - K_{X/Cl}^{sel}) \tag{12}$$

Multiply charged anions are extracted with greater difficulty than the corresponding anions of lower charge, as shown in the series $PO_4^{3-} < HPO_4^{2-} < H_2PO_4^-$ and $SO_4^{2-} < HSO_4^-$. This is the reason why $Bu_4N^+HSO_4^-$ is advantageously used in the preparation of a number of tetrabutylammonium salts; the SO_4^{2-} anion formed by neutralization stays quantitatively in the aqueous phase [9].

An important consequence of the high extraction constants of anions such as I^- and p-$TosO^-$ is that, when they are released during a reaction, they tend to remain preferably bonded to the quaternary cation in the organic phase, thus first decelerating and then stopping the PTC reaction. The obvious conclusion is that a high nucleofugacity is not a sufficient condition for rapid nucleophilic reaction under PTC conditions: it is also necessary that the leaving group is highly hydrophilic: $MeSO_3^-$, Cl^- or Br^- by example give better results than I^-, p-$TosO^-$, etc. [11].

The possibility of extracting highly hydrophilic and highly basic anions such as OH^- and F^- into a low-polarity medium is of particular interest: as pointed out by Dehmlow [37, 57] much of the literature data dealing with extraction constants of OH^- is in error. This is mainly due to two causes: the easy reaction of OH^- with the solvent (in particular $CHCl_3$ and CH_2Cl_2) and the degradation of the quaternary cation Q^+. With highly lipophilic cations, however, such extraction is possible so that the extraction constant of $Hept_4N^+OH^-$ in benzene/water is $\simeq 1$. However, this is 10^4 times lower than the extraction constant of other common anions such as Cl^-, and even very small amounts of other anions present in the solution strongly interfere with the extraction constants. In practice, only a small portion of Q^+X^- extracted into the organic phase is present as Q^+OH^-.

Table 2. Hydration state n of anions associated to quaternary cations in low-polarity solvents in equilibrium with aqueous solutions ($Q^+X^-nH_2O$)

X	$C_{16}H_{33}P^+Bu_3X^-$		$(C_8H_{17})_4N^+X^-$	$(C_8H_{17})_3N^+PrX^-$
	Chlorobenzene [58]	1-Cyanooctane [12]	Toluene [58]	Toluene [58]
N_3	3.0	—	—	—
CN	5.0	5.0	—	—
Cl	3.4	4.0	3.2	2.5
Br	2.1	—	2.4	1.6
I	1.0	—	—	—
SCN	2.0	—	—	—
NO_3	—	0.4	1.5	1.1
SO_4	—	—	18.0	—

Finally, the anions dissolved in the organic phase in the presence of an unmixed aqueous phase are always accompanied by a certain amount of solvation water, and then exist as $Q^+X^-nH_2O$. This may have important consequences on the reaction rate under PTC conditions (see Sect. 2.3.2).

The value of n depends mainly on the nature of the anion, less on the nature of the cation and, according to current knowledge, on the nature of the organic solvent immiscible with water. For most anions associated with quaternary onium cations, n is generally 1 to 5 (Table 2).

With regard to the extraction of organic anions, there seems to have been no systematic study as yet. However, it may be tentatively predicted that small anions, such as formate and acetate, would be extracted with greater difficulty than bulkier anions which, with the same charge, are more lipophilic. More generally, the presence of alkyl chains or lipophilic groups (Hal, NO_2, etc. . . . .) strongly favours the extraction of the anion into the organic phase.

2.2.4 Influence of Solvents

An ideal solvent for PTC must be aprotic and immiscible with water to avoid strong interactions with the ion pairs and thus poor reactivity (see Sect. 2.3.2). Furthermore, it must be chemically stable under the reaction conditions. As shown in Table 3, in the case of $Bu_4N^+Br^-$ taken as standard [25], the extent of extraction largely depends on the solvent polarity. This is the main factor for the choice of the solvent in PTC reactions.

From this point of view, some chlorinated solvents, such as $CHCl_3$ and CH_2Cl_2, appear to be particularly favoured. It should be remembered that in most cases PTC reactions proceed rapidly and thus potentially reactive solvents, e.g. CH_2Cl_2, are not affected by the reaction system in practice.

Polarity is no longer the main factor in the choice of solvent when the ion pair that must be extracted is extremely lipophilic, as in the case of $Bu_4N^+Br^-$, due to a favourable structural combination of both cation and anion. This aspect is of particular interest if we consider that the polarity of the organic medium can both determine the extractability of the ion pair and influence its chemical reactivity. For example, in the substitution reaction of the methanesulfonate group of

Table 3. Apparent extraction constants of tetrabutylammonium bromide [25]

Solvent	$E_{Bu_4N^+Br^-}$ [a]	Solvent	$E_{Bu_4N^+Br^-}$ [a]
CH_2Cl_2	35	trans-ClCH=CHCl	0.1
$CHCl_3$	47	cis-ClCH=CHCl	33
CCl_4	0.1	$ClCH=CCl_2$	0.2
$ClCH_2CH_2Cl$	6.1	CH_3NO_2	168
$ClCH_2CHCl_2$	8.6	$ClCH_2CN$	17,000
$Cl_2CHCHCl_2$	145	$C_2H_5COCH_3$	14
C_6H_5Cl	0.1	$C_2H_5OC_2H_5$	0.1
o-$C_6H_4Cl_2$	0.1	$CH_3COOC_2H_5$	0.2

[a] Calculated from the distribution of 0.1 M $Bu_4N^+Br^-$ between water and the organic solvent

n-octylmethane sulfonate by a series of nucleophiles, the rate constants increase as the polarity of the organic phase decreases [59,60].

Anyway this behaviour is not general and likely depends on the extent of bond making and breaking in the transition state [61] (see Sect. 2.3.2).

In particular cases the presence of cosolvents seems to be somewhat useful. In the generation of dihalocarbenes, the addition of a small amount of ethanol to chloroform in some cases remarkably improves the yield of the corresponding carbene [62]. This is likely due, however, to the formation of a more lipophilic alkoxy anion, which participates in the reaction, rather than to a solvent effect. As a further example, the extraction of tetraalkylammonium hydroxides into aromatic hydrocarbons increases by 1 to 3 orders of magnitude when a small amount of alcohol is added [63]; however, the extracted species is mainly RO^- than OH^- [64].

2.2.5 Salt Effect

The extraction constants reported in the literature are often determined at constant ionic strength of the aqueous phase. However, as the concentration of inorganic salts increases, a salt effect is observed leading to the transfer of organic salts from the aqueous phase into the organic one [65,66].

For example [25], the extraction constants of $Bu_4N^+Cl^-$ and $Bu_4N^+Br^-$ in water-CH_2Cl_2 increase by a factor of 1000 if 2 mol/L of K_2CO_3 is added. HCO_3^- and CO_3^{2-} anions are not extracted under these conditions. A similar salt effect arises from 50% aqueous NaOH used in many current PTC reactions (i.e. the generation of carbanions, carbenes, etc.). This allows the use of quaternary salts which normally show an unfavourable partition coefficient with respect to the organic phase, such as triethyl-benzyl-ammonium chloride (TEBA).

If the aqueous phase is salted, it is also possible to use under PTC conditions solvents partially or fully miscibles with water (i.e. acetonitrile, THF, etc.).

2.3 Mechanism of Phase-Transfer Catalysis

2.3.1 Fundamental Aspects

In his original formulation, Starks [11,12] schematized the mechanism of PTC according to Eq. (2).

The quaternary onium salt transfers the anion from the aqueous phase into the organic one, where the reaction takes place. It then transfers the leaving group into the aqueous phase. This mechanism assumes a partition of the catalyst between the two phases. On the other hand, other conditions being the same, the efficiency of a PTC catalyst is directly related to its solubility in the organic phase [11,12,59,66–68] (see Sect. 2.3.3). The modified scheme 13 may thus be proposed alternatively; in this case, the electroneutrality of the phases is simply maintained by the transport of the anions.

$$R-X + Q^+Y^- \rightarrow R-Y + Q^+X^- \quad \text{(organic phase)}$$
$$\updownarrow \qquad\qquad \updownarrow \qquad\qquad\qquad (13)$$
$$Y^- \;+\; M^+ + X^- \quad \text{(aqueous phase)}$$

It is not possible to make a distinction between the two mechanisms by kinetic determinations [56]. However, the use of liquid membranes [59,67] or indicators [25] has shown that anion exchange occurs at the interface without concomitant transfer of the quaternary cation Q^+, at least in the case of a catalyst completely insoluble in the aqueous phase (Eq. (13)).

A series of tests indicates that the reaction takes place in the organic phase and thus governs the process as a whole. In particular, in the simplest case of an irreversible nucleophilic substitution (Eqs. (2) and (13)) and in the presence of an excess of the anionic reagent in the aqueous phase, the reaction follows pseudo-first-order kinetics (Eq. (14)). The observed rate constants depend linearly on the concentration of the catalyst in the organic phase (Eq. (15)) [12,59,68].

$$\text{rate} = k_{obs}[RX] \qquad (14)$$
$$k_{obs} = k[Q^+Y^-]_{org} \qquad (15)$$

Typical interfacial reaction rates depend on the stirring speed [69]. As a rule, under PTC conditions, observed rate constants are independent of the stirring speed above 250–300 rpm [11,12,59,66]. This demonstrates that interfacial phenomena are not important for this kind of reactions.

The similarity between activation parameters measured under both anhydrous homogeneous conditions and PTC conditions in the case of nucleophilic substitutions [68] confirms that rates are not controlled by diffusion.

While kinetic aspects of PTC have been examined in detail for specific reactions [12,59,66], other works have developed a more general theory of PTC kinetics, by analyzing a number of aspects and variations [25,36,56,65].

Kinetic data exclude the possibility of the reaction occurring in micellar phase. Even the possibility of inverted micelles interference [70] can be excluded. Aggregation numbers from 1 to 15 were found for a series of quaternary ammonium and phosphonium salts [12,60,71], in low-polarity anhydrous solvents and within concentration ranges comparable with those utilized in the kinetic measurements and synthetic processes. The linear realtionship between pseudo-first-order rate constants and catalyst concentration within a large range (Eq. (15)) indicates that a limited number of water molecules associated with quaternary salts are sufficient to hold the formation of aggregates to a level which as no influence on the reaction rates [59].

2.3.2 Anionic Reactivity of Quaternary Onium Salts

According to Coulomb's law, cation-anion interactions can be minimized by increasing either the dielectric constant of the medium or the distance between the ions. On the basis of Eq. (16) the interaction energy between cation and anion may be calculated (e is the charge of the electron, ε the

$$E_{\pm} = e^2 N/\varepsilon r = 33.18/\varepsilon r \text{ kcal/mol} \tag{16}$$

dielectric constant, r the distance between the centers of cation and anion expressed in Å, N Avogadro's number). By this method it has been calculated that, for K^+Br^- and $Bu_4N^+Br^-$, these energies are 4.5 and 2.4 kcal/mol in a solvent with a low dielectric constant such as dioxane ($\varepsilon = 2.2$), and are 0.26 and 0.13 kcal/mol in a solvent with a high dielectric constant such as acetonitrile ($\varepsilon = 39$) [72].

It may be assumed that with decreasing cation-anion interactions the free energy of activation of a displacement reaction will correspondingly decrease. This prediction is in agreement with a series of data [73]. For example, Table 4 reports the rate constants for nucleophilic displacement of halide in 1-bromobutane by K^+ and Bu_4N^+ phenoxides in dioxane and acetonitrile. The reactions of $Bu_4N^+PhO^-$ always proceed more rapidly than the corresponding reactions involving K^+PhO^-.

With K^+PhO^- the rates strongly depend on the dielectric constant of the solvent, but with $Bu_4N^+PhO^-$ the rates are largely independent of dielectric constant. In DMF the rates for both salts are nearly the same.

We can therefore conclude that reaction rates are extremely sensible both to the dielectric constant of the medium and to the ionic radius of the cation. The effect of the dielectric constant is very pronounced when the ionic radius is small, and the effect of the ionic radius is strongest when the dielectric constant value is small.

Conductivity measurements demonstrated that both K^+ and Bu_4N^+ phenoxides exist practically as an ion pair in dioxane while they are largely dissociated in acetonitrile [74]. Therefore, the fact that quaternary ammonium phenoxide in dioxane is more reactive than the corresponding potassium salt is mainly due to the greater interionic distance and, accordingly, to the lower cation-anion interaction energy.

The use of quaternary cations with at least one long alkyl chain ensures sufficient solubility of the salt even in very low dielectric-constant organic solvents, leading to the formation of an ideal highly reactive ion pair.

Table 4. Influence of interionic distance (r) and dielectric constant (ε) on reaction rates of ion pairs [73]

$n\text{-}C_4H_9Br + PhO^-M^+ \rightarrow PhOC_4H_9\text{-}n + M^+Br^-$

Solvent	ε	$k \times 10^5$ (mol^{-1} s^{-1})		$k_{Bu_4N^+}/k_{K^+}$
		PhO$^-$K$^+$	PhO$^-$Bu$_4$N$^+$	
dioxane	2.2	0.01	330	33,000
acetonitrile	39	40	300	7.5

Quaternary salts exist mainly as dissociated ions in protic solvents, as ion pairs in equilibrium with dissociated species in dipolar aprotic solvents, and as ion pairs in equilibrium with quadrupoles or more complex aggregates in low-polarity solvents [1,12,60,75-79]. Under PTC conditions, it is highly unlikely that reactivity is due to free anions since the reaction occurs in the apolar organic phase where their concentration is virtually null.

Many authors [12,25,59,60,66,80,81] independently arrived at this conclusion by kinetic measurements. As noted above, within a wide concentration range, observed rate constants are directly proportional to the concentration of Q^+Y^-. The Q^+Y^- ion pair must therefore be the dominant nucleophile. If, on the other hand, the free anion is the reactive species, rates would be proportional to $\sqrt{[Q^+Y^-]}$ in the organic phase [25,80,81].

Of further significance both mechanistically and practically is the comparison between rate constants under PTC and homogeneous conditions, particularly in dipolar aprotic solvents.

In the nucleophilic displacement of n-octyl methanesulfonate (Eq. (17)) taken as standard substrate, by a series of anions, a rather restricted reactivity range in the following order was found: $N_3^- > CN^- > Br^- \simeq I^- > Cl^- > SCN^-$. This is inconsistent with the behaviour in both protic and dipolar aprotic solvents [59]. This behaviour is largely due to the specific solvation of anions by a limited number of water molecules in the organic phase (see Table 2).

$$n\text{-}C_8H_{17}OSO_2Me + Y^- \xrightarrow[\text{PhCl}-H_2O]{Q^+Y^-} n\text{-}C_8H_{17}Y + MeSO_3^- \qquad (17)$$

By repeating the same reactions in homogeneous solution in apolar organic medium, it was found that both absolute and relative rates are dramatically modified. The absolute rates increase to a maximum of one order of magnitude, the values becoming higher with increasing electronegativity and decreasing polarizability of anions. When the reactions are carried on under homogeneous conditions in the presence of small amounts of water sufficient to regenerate the hydration sphere of the anions in the organic phase, the second-order rate constants found are equal to those obtained under PTC conditions [59] (Table 5). The anionic reactivity sequence in anhydrous homogeneous apolar solvents is $CN^- > N_3^- > Cl^- > Br^- > I^- > SCN^-$, i.e. the same as that in dipolar aprotic solvents. Second-order rate constants measured for the indicated series of anions in chlorobenzene-water under PTC conditions are very similar to those obtained in anhydrous DMSO homogeneous solution. The rates also depend on the polarity of the organic solvent used under PTC conditions and, in the case of the test reaction with the $MeSO_3^-$ leaving group, increase as the polarity and/or polarizability of the solvent decrease. The range is small (a factor of 3.4 passing from benzonitrile to cyclohexane), but the trend is the same as that observed under anhydrous homogeneous conditions [59] (Table 6), although the dependence of the reaction rate on the nature of the solvent is considerably greater under homogeneous conditions.

These results indicate that second-order rate constants under PTC and dipolar aprotic solvents conditions are very close. The observed rate constants under PTC conditions obviously depend on the amount of catalyst: it is generally used in 1 to

Table 5. Second-order rate constants[a] for the reaction of n-octyl methanesulfonate with various nucleophiles (Y⁻)[b] under PTC conditions, and in anhydrous and wet homogeneous organic solutions [59)]

Y	PTC conditions (PhCl-H$_2$O) $k \times 10^3$ (mol^{-1} s^{-1})[c]	Homogeneous conditions $k \times 10^3$ (mol^{-1} s^{-1})[c]		
		Anhydrous PhCl	Wet PhCl	Anhydrous DMSO
N$_3$	19.1 (6.8)	70.4 (23.5)	19.6	13.5
CN	11.7 (4.2)	86.7 (29.0)	13.3	33.8
Cl	1.8 (0.6)	19.7 (6.6)	2.2	3.6
Br	3.2 (1.1)	8.1 (2,7)	3.3	2.3
I	2.8 (1.0)	3.0 (1.0)	2.7	0.5
SCN	0.5 (0.2)	0.75 (0.3)	0.5	0.3

[a] At 60 °C; [b] From C$_{16}$H$_{33}$P$^+$Bu$_3$Y$^-$; [c] Relative rates in parentheses

Table 6. Solvent effect on the rate of the reaction of n-octyl methanesulfonate with bromide ion[a] under PTC and anhydrous homogeneous conditions [59)]

Organic solvent	PTC conditions		Homogeneous conditions	
	$k \times 10^3$ (mol^{-1} s^{-1})[b]	k_{rel}	$k \times 10^3$ (mol^{-1} s^{-1})[b]	k_{rel}
Cyclohexane	5.5	3.4	40.5	9.0
Toluene	3.7	2.3	10.7	2.4
Chlorobenzene	3.2	2.0	8.1	1.8
o-Dichlorobenzene	3.2	2.0	5.5	1.2
Benzonitrile	1.6	1.0	4.5	1.0
DMSO	—	—	2.3	0.5
Methanol	—	—	0.2	0.05

[a] From C$_{16}$H$_{33}$P$^+$Bu$_3$Br$^-$; [b] At 60 °C

5% molar ratio with respect to the substrate, and reaction times are correspondingly longer. This disadvantage, however, is balanced by the possibility of operating in the absence of solvents, with the substrate and/or the reaction product acting as organic phase.

2.3.3 Dependence of Anionic Reactivity on the Structure of the Onium Cation

Various authors have proposed different empirical evaluations of anionic reactivity under PTC conditions based on the structure of the catalyst. In most cases, however, comparison of the results is very difficult, since a number of factors are involved simultaneously and non-selectively, often counteracting each other. As previously discussed, fundamental requirements for catalytic efficiency under PTC conditions are: i) the lipophilicity of the catalyst; ii) the extraction selectivity; iii) the cation-anion separation within the ion pair; iv) the poor solvation of the anion. Other factors, however, may assume great importance, depending both on the experimental conditions and on the kind of reaction, i.e. on the structure of the substrate and of the anion. As shown in Table 7, in the case of a typical aliphatic nucleophilic displace-

Table 7. Influence of the nature of quaternary cation (Q^+) on the rate constants of the reaction of n-octyl methanesulfonate with bromide ion under PTC conditions [59,67]

Q^+	% Catalyst in the organic phase	$k_{obs} \times 10^5$ [a] (s^{-1})	$k \times 10^3$ [b] $(mol^{-1} s^{-1})$
$(C_8H_{17})_4N^+$	100	20.4	5.1
$C_{16}H_{33}N^+Bu_3$	100	17.3	4.3
$C_{16}H_{33}N^+Pr_3$	94	15.9	4.0
Bu_4N^+	83	12.0	3.6
Pr_4N^+	2.5	0.2	2.4
$PhCH_2N^+Bu_3$	95	8.2	2.1
$PhCH_2N^+Pr_3$	17	1.3	1.9
$C_{16}H_{33}P^+Bu_3$	100	12.8	3.6
Bu_4P^+	97	10.4	2.7

[a] In PhCl-H_2O at 60 °C; [b] $k = k_{obs}/[Q^+Br^-]_{org}$

ment [59,67], the observed pseudo-first-order rate constants differ by up to two orders of magnitude, depending on the structure of the catalyst. Second-order rate constants, however, considering the actual concentration of the catalyst in the organic phase, remain within a narrow range and differ by a maximum factor of 2.5. The lipophilicity of the cation is thus fundamentally important while the structural variations of the latter are less significant. It can, however, be noted that the anionic reactivity rises as the steric hindrance around the cationic center increases, in agreement with results discussed in Sect. 2.3.2. This can be achieved both by increasing the length of alkyl chains and by decreasing the length of the heteroatom-carbon bond, as observed passing from phosphonium to ammonium salts.

When all or most of the alkyl chains bonded to the quaternary salt are methyls or ethyls, the catalyst is no longer suitable for PTC. In this case, the salt is either exceedingly hydrophilic or, when one long chain and three short groups are bonded to the cationic center, shows surfactant properties, yielding stable emulsions between the aqueous and organic phases [11,59].

TEBA is an interesting case. It is normally used in generations and subsequent reactions of carbanions and carbenes; its catalytic efficiency is mainly due to the "salting-out" effect of the aqueous phase, since the latter is normally 50% NaOH. On the other hand, it shows very low catalytic activity in nucleophilic displacements [36,66].

Anionic reactivity as a function of the length of the alkyl chains bonded to the cationic center is shown in Table 8 [59]. In the series of hexadecyl-trialkylammonium salts the major increase in reactivity occurs when passing from the trimethyl to the triethyl derivative; with higher homologous, the results are nearly the same. Determinations for methyl and ethyl derivatives were obviously limited to anhydrous homogeneous solutions, since these salts behave as typical surfactants under PTC conditions.

A systematic study of reactivity variations as a function of heteroatom-carbon bond length has not yet been performed, limited data being available only for R_4N^+ and R_4P^+ cations (see Tables 7 and 8).

The most important variation due to the nature of the heteroatom is the chemical

Table 8. Influence of the nature of quaternary cation (Q^+) on the rate constants of the reaction of n-octyl methanesulfonate with bromide ion in anhydrous chlorobenzene and DMSO [59]

Q^+	PhCl	DMSO
	$k \times 10^3$ (mol^{-1} s^{-1})$^{a, b}$	$k \times 10^3$ (mol^{-1} s^{-1})$^{a, b}$
$C_{16}H_{33}P^+Bu_3$	8.7	3.5
Bu_4P^+	9.1	—
Bu_4N^+	20.2	3.2
$(C_8H_{17})_4N^+$	19.6	—
$C_{16}H_{33}N^+Bu_3$	18.2	3.2
$C_{16}H_{33}N^+Pr_3$	19.2	3.5
$C_{16}H_{33}N^+Et_3$	18.6	2.9
$C_{16}H_{33}N^+Et_2Me$	8.3	—
$C_{16}H_{33}N^+EtMe_2$	6.5	—
$C_{16}H_{33}N^+Me_3$	3.9	3.2

^{a} At 60 °C; ^{b} [Q^+Br^-] = 0.50–0.58 × 10^{-2} M

stability of the salt. Phosphonium salts are generally much more thermally stable than the corresponding ammonium salts [11,36] (up to 150–170 °C when associated with anions such as Cl$^-$, Br$^-$ or CN$^-$). Ammonium salts, on the other hand, lose their reactivity when heated to 110–120 °C. While this difference may not seem particularly important, since in most cases reactions under PTC conditions proceed at temperatures below 110 °C, it becomes significant in those cases requiring longer reaction times or higher temperatures under pressure. However, phosphonium salts are severely affected by OH$^-$, irreversibly yielding the corresponding phosphine oxides at temperatures of 50 °C or higher [36,82–85].

No stability data on arsonium salts are available. These salts have been used, from time to time, by several authors as an alternative to phosphonium salts, apparently with comparable reactivity.

Sulfonium salts have occasionally been indicated as potential PTC catalysts; their use has been claimed in a patent [86]. A recent systematic study of these salts, however, showed that, while they are relatively stable in polar protic solvents [87], they are rapidly decomposed to sulfides when dissolved in aprotic apolar solvents [88,89]. This clearly indicates that they cannot be used as PTC catalysts [89]. Some of them have, however, been employed in micellar reactions [90–92].

Alkylephedrinium and more generally β-hydroxyethylammonium salts show a particularly high activity in the reduction of carbonyl groups by borohydride ions [93,94]. A tentative explanation is that the hydroxy group of the catalyst may form a hydrogen bond with the carbonyl group, activating it toward the nucleophilic attack of BH_4^-. On the other hand, these particular catalysts show a lower activity than normal quaternary salts in other typical PTC reactions (see Sect. 7.9.1).

2.3.4 Mechanism in the Presence of Concentrated Aqueous Bases

A great number of reactions require the presence of a strong base, usually a concentrated alkaline aqueous solution, to generate reactive anions. Examples are C-/O-alkylations of carbanions, H/D exchanges, formation of carbenes via α-elimination, β-eliminations, nucleophilic additions, isomerizations, etc. Relatively strong acids, such as 1,3-dioxo derivatives, are dissolved by aqueous NaOH and can be extracted by cation Q^+ as ion pair into the organic phase where the anion reacts, for example, with an alkylating agent. The mechanism is therefore the usual one (Eqs. (2) and (13)).

Even in the case of relatively weak acids such as alcohols ($pK_a \simeq 18$) the mechanism is essentially the same. Although in the aqueous phase equilibrium (18) is largely shifted to the left, the high hydrophilicity of OH^- leads

$$OH^- + ROH \rightleftharpoons RO^- + H_2O \tag{18}$$

to a very favoured extraction into the organic phase of the Q^+OR^- ion pair rather than Q^+OH^- [64].

The situation is quite different for weaker organic acids with pK_a in the range 22 to 25 [95]. In this case, proton abstraction is likely to occur at the interface [7,19]. Makosza showed that generation and alkylation of carbanions in the presence of aqueous 50% NaOH and of a non-miscible organic phase may occur even in the absence of PTC catalysts, provided that the alkylating agent is highly reactive. For example, $PhCH_2CN$ is alkylated by 1-iodobutane and 50% aqueous NaOH [96] by stirring the mixture at 80 °C. Under these conditions, the concentrations of $PhCH_2CN$ in the aqueous phase and of the sodium salt of the carbanion in the organic phase were less than 2 and 5 ppm, respectively.

In these reactions the quaternary salt transports the carbanion from the interface into the organic phase where the reaction occurs. This was proved as follows: i) the deeply coloured 9-fluorenyl anion was transported into the organic phase from the interface by equilibrating an aqueous 50% NaOH solution with a solution of fluorene in benzene in the presence of $Hex_4N^+Cl^-$; ii) stirring rates must be at least 750–800 rpm to obtain reproducible results in the generation of dihalocarbenes starting from chloroform and 50% NaOH in the presence of a quaternary salt [97]; this is consistent with the typical behaviour of interfacial reactions [69], since usual PTC reactions proceed at a constant rate over 250–300 rpm (see Sect. 2.3.1).

The same behaviour is observed for the other important group of reactions involving the generation of halocarbenes via halocarbanions (Eqs. (19)–(21)).

$$HCCl_{3\,(org)} + OH^-_{(aq)} \rightleftharpoons CCl^-_{3\,(interphase)} + H_2O \tag{19}$$

$$CCl^-_{3\,(interphase)} + Q^+X^-_{(aq,\,org)} \rightleftharpoons Q^+CCl^-_{3\,(org)} + X^-_{(aq)} \tag{20}$$

$$Q^+CCl^-_{3\,(org)} \rightleftharpoons Q^+Cl^-_{(org)} + :CCl_{2\,(org)} \tag{21}$$

The formation of carbene (Eq. (21)) is a reversible process [98], but it irreversibly reacts with the substrate in the organic phase to result in additions, insertions, etc. (see Sect. 7.5).

The extreme interest in this kind of reaction, which furnishes carbenes by an easily accessible process and has practically supplanted all the previously used methods, is also due to the fact that side reactions, and in particular irreversible hydrolysis, are largely excluded by the total absence of OH^- and the extremely small amount of water in the organic phase.

3 Ion Pair Extraction

At the same time as the classic liquid-liquid PTC was reported, Brändström introduced and developed a complementary technique [9] involving the use of lipophilic quaternary onium salts in stoichiometric amounts instead of catalytic amounts. For this reason, PTC is the method of choice in most reactions. "Ion pair extraction", however, is particularly useful in the case of leaving groups with both high nucleofugacity and high lipophilicity, such as I^- and $ArSO_3^-$. Brändström's technique consists of extracting the anionic reactant into the organic phase from the aqueous solution using quaternary salts of highly hydrophilic anions such as HSO_4^-. In this way, the desired quaternary salts are prepared and the reaction can be rapidly performed with an equimolar amount of substrate in a homogeneous apolar organic solution. The disadvantage of using stoichiometric amounts of quaternary cation, generally Bu_4N^+, is balanced by the possibility of using substrates with highly reactive leaving groups. An additional advantage is that strictly anhydrous conditions may be used. An example is the facile synthesis of diborane by the reaction of $Bu_4N^+BH_4^-$ with simple alkyl halides [99].

Major applications of ion pair extraction are: alkylations (mainly C-alkylations), synthesis of esters, halogen and pseudohalogen exchanges, oxidations and reductions [9]. In these cases, it is obviously important to recover and recycle the quaternary salt. A number of methods to this end were proposed by Brändström [9].

4 Solid-Liquid and Solid-Gas Phase-Transfer Catalysis

Classic PTC requires the presence of an aqueous-organic two-phase solution. As discussed previously, the catalyst (a quaternary salt, a macrocyclic or macrobicyclic polyether as a complex, and in general whatever kind of PTC catalyst, see Sect. 5) extracts some water into the apolar organic phase, mostly as specific solvation sphere of the anion. The amount of water transported, though scare, may affect the reaction to give undesired side effects.

These can be avoided by working with a solid anionic reagent suspended in an anhydrous organic solution of the substrate. Under these conditions, the PTC catalysts transfer the anion from the surface of the crystal lattice to the organic phase. A typical example is the synthesis of dichlorocarbene by thermal decomposition of Cl_3C-COONa (Eq. (22)) suspended in anhydrous $CHCl_3$ in the presence of catalytic amounts of a quaternary salt.

$$Cl_3C\text{-COONa} \rightarrow NaCl + CO_2 + :CCl_2 . \tag{22}$$

The generated carbene can be trapped by an alkene, furnishing the corresponding dichlorocyclopropane in very high yields [100, 101]. Under liquid-liquid PTC conditions, a different reaction occurs, the presence of water stopping the decomposition of Cl_3C-COONa at the level of $CHCl_3$ (Eq. 23).

$$Cl_3C\text{-COONa} + H_2O \rightarrow NaHCO_3 + CHCl_3 . \tag{23}$$

The first step of the classical Gabriel synthesis (N-alkylation of phthalimide) gave excellent results under solid-liquid PTC conditions [102]. Lower yields are obtained under liquid-liquid PTC conditions, due to hydrolytic reactions.

The generation of carbanions and subsequent reactions have been analogously carried out under solid-liquid PTC conditions, in the presence of both crown ethers (see Sect. 5.1) and quaternary ammonium salts. The highly satisfactory yields obtained in these cases are mainly due to the possibility of suppressing side reactions [103]. A negative aspect of solid-liquid PTC is that the solution of the crystal is in most cases the rate-determining step of the process. As a further complication the salt generated during the course of the process can cover the surface of the crystals, thus slowing down or stopping the reaction. This alternative technique, while in principle as interesting as liquid-liquid PTC, has not yet been completely examined in a comparative study. Dipolar aprotic solvents like acetonitrile have been often used to increase the solubility of the salt [104, 105]. The anion-dependent different catalytic activity of quaternary salts and crown ethers (see Sect. 5.1) may be due to a balance between the greater effectiveness of crown ethers in dissolving salts and their relatively small complexation constant. This constant, however, varies widely, depending on the anion. Finally, mechanical factors, i.e. pulverization of the salt, stirring rate, etc., must be taken into account.

A further variation of classical PTC is gas-solid PTC. In this technique a continuous flow of the gaseous substrate passes through a fixed bed of the reacting salt and of a thermally stable catalyst, normally a phosphonium salt, eventually adsorbed on silicagel [106, 107]. This technique has been applied to simple nucleophilic displacements such as the synthesis of halides and alkyl esters (Eq. (24)) or the halogen exchange between

$$R-X_{gas} + M^+Y^-_{solid} \xrightarrow{Q^+X^-} R-Y_{gas} + M^+X^-_{solid} \quad (24)$$
$$X = Cl, Br; \quad Y = Br, I, RCO_2$$

different alkyl halides [108].

Even carbowax (a chemically and thermally stable poly(ethylene glycol)), when adsorbed on an inorganic salt with no other solid support, may act as a very efficient gas-solid phase-transfer catalyst. This system has been employed, for example, in the Williamson synthesis of ethers and thioethers, starting from alkyl halides and phenols or thiols in the presence of potassium carbonate as a base [109].

Gas-solid PTC shows the advantage that pure products are obtained directly, due to the absence of aqueous and organic solvents.

5 Other PTC Catalysts

5.1 Crown Ethers

Pedersen's discovery of macrocyclic polyethers (crown ethers) [2, 3] represents one of the most significant innovations in organic chemistry in recent years: specific complexation of the cation by the macrocyclic ligand both facilitates the solubility of salts in low-polarity solvents and furnishes non-solvated highly reactive anions by increasing the dissociation of ion pairs.

The high ratio between the number of heteroatoms (normally oxygen) and carbon atoms makes unsubstituted crown ethers highly hygroscopic and water-soluble. Their solubility is strongly affected by the presence of aromatic rings or alkyl chains. Compounds fused with aromatic rings, particularly those bearing more than one benzo group, are nearly insoluble in water and scarcely soluble in alcohols and most organic solvents at room temperature. They are very soluble in CH_2Cl_2 and $CHCl_3$. Upon hydrogenation of the aromatic rings, crown ethers show a remarkable increase in solubility in organic solvents, even those of low polarity; at the same time, solubility in water decreases [3,40].

Two conditions should be satisfied to allow the use of crown ethers as catalysts under PTC conditions [110,111]: i) a favourable partition ratio of complexed crown ether in the organic phase with respect to the aqueous one; ii) a high complexation constant between the ligand and the salt in the organic phase.

The other conditions being the same, the more lipophilic the ligand, the higher the unsolvated anion concentration in the organic phase. This, however, is also due to the stability constant of the complex as well as to: i) the number and nature of ligand atoms in the ring; ii) the relative sizes of the cation and the macrocycle cavity; iii) the degree of solvation of both anion and cation-crown complex; iv) the effect of steric and conformational factors in the ring.

The stability of the cation-ligand complex obviously plays a fundamental role in the solubilization of the salt; the anion too, however, must be considered from this point of view. The complexed cation, being surrounded by organophilic groups, can be extracted into the apolar medium with relative ease. For anions this may not be true since salts of "hard" anions, such as F^- or SO_4^{2-}, are generally not easily transferred into the organic phase by lipophilic cyclic polyethers, while "soft" ones, such as I^- and SCN^-, are readily extracted. This behaviour is consistent with that observed in the case of the solubilization of salts in an organic medium from the crystalline state [3].

Crown ethers, such as *1–3*, have been used as PTC catalysts [38–40,43–47], especially in solid-liquid phase-transfer. The applications of lipophilic crown ethers in the more classic liquid-liquid PTC are limited [112–116]. Only recently quantitative data on anionic reactivity under these conditions have been reported [110,111,117] and compared with those of other catalysts. Two are the reasons of this fact: the higher cost of crown ethers with respect to quaternary salts, and especially the underestimation of

1a, R=H
1b, R=$C_{14}H_{29}$-n
1c, R=$CH_2OC_{16}H_{33}$-n
1d, R=CH_2X

2a, R=H
2b, R=alkyl

3

the importance of lipophilicity and complexation constants. This importance has not been fully recognized, even in a very recent work [111].

The readily available perhydrodibenzo[18]crown-6(PHDB) *3* in an aqueous organic two-phase system is quantitatively dissolved in the organic phase, either as such or as a complex. Kinetic determinations showed [110] that, in the presence of catalytic amounts of *3*, nucleophilic displacements under two-phase conditions follow the classical PTC mechanism, and the observed pseudo-first-order rate constants are linearly correlated with the amount of complexed crown ether in the organic phase. A study of reaction (25) with a series of anionic nucleophiles revealed a relatively restricted reactivity range and the following sequence of relative rates:

$$N_3^- > I^- \simeq Br^- > CN^- > Cl^- > SCN^-.$$

These results are consistent with those obtained under similar conditions with quaternary salts. Second-order rate constants are also very similar in both cases (Table 9). However, the catalytic efficiencies of *3* and quaternary salts differ greatly,

$$n\text{-}C_8H_{17}OSO_2Me_{org} + Y^-_{aq} \xrightarrow{[K^+ \subset PHDB]Y^-} n\text{-}C_8H_{17}Y_{org} + MeSO_3^-{}_{aq} \quad (25)$$

since both observed rates depend on the actual concentration of the anion in the organic phase and thus, for complexed crown ethers, on their stability constants. When all the other conditions are the same, the constants strongly depend on the nature of the anion, with high or low values for "soft" or "hard" anions (see above). Table 9 shows this quantitatively: the ratio $k_{obs}^{PHDB}/k_{obs}^{Q^+}$ ($Q^+ = C_{16}H_{33}P^+Bu_3$) is in the range 0.02 (in the case of Cl^-) to 2.4 (in the case of I^-). As a comparison, the crown ether shows a 2% and 92.8% complexation in the cases of Cl^- and I^-, respectively, in a water-chlorobenzene two-phase system. Furthermore, in the case of "soft" anions such as I^-, second-order rate constants are 1.5 to 2.5 times higher than those obtained with quaternary salts.

It can be concluded that crown ethers are not generally the catalysts of choice for

Table 9. Reaction of n-octyl methanesulfonate with various nucleophiles (Y^-) under PTC conditions in the presence of perhydrodibenzo[18]crown-6(PHDB) or hexadecyl-tributylphosphonium salts (Q^+Y^-) [110]

Y	PHDB · KY			Q^+Y^-		
	k_{obs} (s^{-1})[a]	% Complexation	$k \times 10^2$ [b,c] (mol^{-1} s^{-1})	k_{obs} (s^{-1})[a]	$k \times 10^2$ [b] (mol^{-1} s^{-1})	$k_{obs}^{PHDB}/k_{obs}^{Q^+}$
N_3	25.1	13.3	5.1 (2.8)	188	4.7	0.13
CN	3.5	12.2	0.72 (0.4)	120	3.0	0.03
Cl	0.3	2.0	0.38 (0.2)	17	0.44	0.02
Br	10.7	16.8	1.6 (0.9)	29.6	0.74	0.4
I	66.8	92.8	1.8 (1.0)	27.6	0.69	2.4
SCN	6.6	83	0.2 (0.1)	5.2	0.13	1.3

[a] At 70 °C; [b] $k = k_{obs}/$[complexed PHDB], or $k = k_{obs}/[Q^+Y^-]$; [c] Relative rates in parentheses

liquid-liquid PTC [110,111]. Lipophilic crown ethers can be used as efficient catalysts only in the presence of anions with a dispersed charge and/or being easily polarizable. A convenient path to synthesize alkyl-substituted crown ethers has been recently described [118].

As in the case of quaternary salts also hydrophilic crown ethers (e.g. [18]crown-6, *1a*), when complexed with salts of highly lipophilic anions in strongly alkaline solutions, show a partition coefficient which is more favourable for the organic phase, and operate as efficient catalysts under liquid-liquid phase-transfer conditions [119,120].

Identical results are obtained in the determination of the hydration number of the anion associated either to the cation complexed by a crown ether or to a quaternary salt. In both cases, the hydration sphere of the anion in the organic phase is the cause of the anomalous nucleophilicity scale found under PTC conditions [110].

The main use of crown ethers in PTC is in solid-liquid phase-transfer. In particular, it has been emphasized [38] that they should be the catalysts of choice under such conditions. Due to its particular structure, the crown ether can approach the crystalline lattice so that the extraction and subsequent complexation of the cation require very little cation displacement, while the anion is contemporaneously associated to the complex. In the case of quaternary salts, on the other hand, the steric hindrance around the cationic center makes its interaction with the surface of the crystal difficult and the solution mechanism more complicated: in this case only the anion must be extracted to displace the one originally associated with the lipophilic cation. These conclusions met with some scepticism [121]. Indeed, onium salts, cryptands, polypodes, polyamines, etc. have been successfully employed as solid-liquid PTC catalysts (see Sects. 4, 5.2, 5.3, 5.4). Moreover, when the catalytic activity of quaternary salts, crown ethers and polyamines was compared with respect to the extraction of anions from the crystalline state into an organic solvent, crown ethers were found to be the best system for the transport of CN^-. The catalytic effectiveness is completely reversed in the case of other anions, such as F^- and CH_3COO^-, the quaternary salt being the most efficient in these cases [122].

5.2 Cryptands

Azamacrobicyclic ethers (cryptands) are of particular interest for anionic activation: they afford stable inclusion complexes of the cation (cryptates), in which the cation is surrounded by a spherical ligand 10 Å in size which takes the place of the solvation sphere [5,41,42]. In cryptates the interaction between ions is thus minimized. High anion reactivity may moreover be expected in apolar medium [41,42]. The stability and selectivity of the complexation with respect to the topology of the ligand have been extensively studied [5,41,42], but anionic activation by cryptates [41,42] and their use as liquid-liquid [116,123-129] or solid-liquid [42,123] PTC catalysts have been hardly treated at all. In this case considerations advanced for crown ethers also apply: simple cryptands, e.g. *4a*, show low solubility in apolar organic medium and very high solubility in water. Most of lipophilic cryptands, e.g. *4b* and *5*, must be appropriately synthetized to be used for PTC [124,127,128,130,131]. Other cryptands, functionalized with particular groups, to allow their condensation with lipophilic chains, e.g. *4c, d*, have recently been described [116].

Table 10. Second-order rate constants for the reaction of *n*-octyl methanesulfonate with various nucleophiles (Y⁻) under PTC conditions and in anhydrous and wet chlorobenzene [132]

Y	PTC conditions (PhCl-H$_2$O)				Homogeneous conditions					
					Anhydrous PhCl				Wet PhCl	
	$k \times 10^3$ [a] $(mol^{-1} s^{-1})$	k_{rel}	Hydration state n of $[K^+ \subset (2.2.2, C_{14})] Y^-$ nH_2O	% Complexation	$k \times 10^3$ [a] $(mol^{-1} s^{-1})$	k_{rel}	% Complexation		$k \times 10^3$ [a] $(mol^{-1} s^{-1})$	
N$_3$	28.6	6.1	4.8	95	150.0	17.3	90		24.5	
CN	15.5	3.0	7.2	97	183.3	21.1	45		15.0	
Cl	3.5	0.7	5.3	95	50.8	4.8	43		6.5	
Br	5.6	1.2	4.3	97	36.6	4.2	97		6.8	
I	4.1	1.0	3.2	100	8.7	1.0	100		5.0	
SCN	1.1	0.23	2.8	100	1.5	0.17	100		0.95	

[a] At 60 °C

Complete kinetic determinations are apparently limited to the lipophilic [2.2.2, C_{14}] cryptand 4b [132], which behaves as a PTC catalyst. In this case as well, reactions follow a mechanism similar to that of quaternary salts: the observed pseudo-first-order rate constants are linearly correlated with the concentration of the cryptate in the organic phase. Moreover, in a water-chlorobenzene system it is found entirely in the organic phase, as cryptand and/or cryptate. Anion exchange does not require the simultaneous transfer from the aqueous phase of the cationic partner.

The reactions can be simulated in homogeneous organic solution by adding the water associated to the cryptate and substrate to the two-phase system, and give a reactivity scale of

$$N_3^- > CN^- > Br^- \simeq I^- > Cl^- > SCN^-.$$

The hydration state of anions in cryptates $[K^+ \subset (2.2.2, C_{14})]Y^-$ is the same as that in quaternary salts. Second-order rate constants (Table 10) are 1.5 to 2.6 fold greater than those measured under the same conditions in the case of quaternary salts. From a certain point of view, the results are similar to those for crown ethers, although cryptates show noticeably higher complexation constants than crown-ethers (cryptate effect [5]). In this case catalyst 4b is in the organic phase and nearly quantitatively complexed. In other words, the ratio between the pseudo-first-order rate constants of cryptates and quaternary salts is equal to that between their second-order rate constants, with the former depending on the actual concentration of the catalyst in the organic phase.

Cryptates represent the best actual approach to a model of solvent-separated ion pairs and, in an apolar solvent, the weak cation-anion interaction and the scarce solvation allow an approach to gas-phase chemistry under liquid conditions [41, 42, 133].

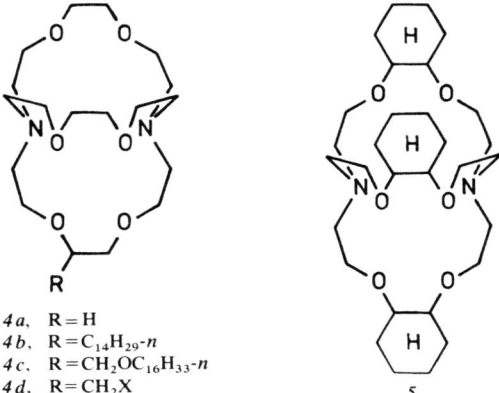

4a, R=H
4b, R=$C_{14}H_{29}$-n
4c, R=$CH_2OC_{16}H_{33}$-n
4d, R=CH_2X

5

The simple [2.2.2] system 4a probably shows optimum separation between the metal cation and the anion [41]. In an effort to achieve a more efficient separation, perhydrotribenzo[2.2.2]cryptand (2.2.2, PHTB) 5 was used. Anyway the second-order rate constants obtained were nearly the same as found for 4b: the three cyclohexane rings only make the system more rigid, causing a drastic decrease of the complexation constants (Table 11) [134].

Recent kinetic data [111] confirm that, under PTC conditions, cryptands are more active catalysts than crown ethers and quaternary salts.

Table 11. Second-order rate constants for the reaction of n-octyl methanesulfonate with various nucleophiles (Y^-) under PTC conditions and in anhydrous organic solution in the presence of (2.2.2, PHTB) cryptand [134]

Y	PTC conditions (PhCl-H_2O)			Homogeneous conditions (anhydrous PhCl)		
	$k \times 10^2$ [a] (mol^{-1} s^{-1})	k_{rel}	% Complexation	$k \times 10^2$ [a] (mol^{-1} s^{-1})	k_{rel}	% Complexation
Cl	3.8	0.8	9	—	—	3
Br	5.5	1.2	42	37.4	3.6	10
I	4.5	1.0	89	10.5	1.0	93
SCN	0.8	0.2	92	2.3	0.19	100

[a] At 60 °C

5.3 Podands and Polypodands [135]

Open-chain polyethers (podands) have been used as catalysts both for anion activation in homogeneous medium and under PTC conditions [47, 111, 136–138]. Variations of the *endo*-polarophilicity/*exo*-lipophilicity ratio, the stability of complexes, and the selectivity with respect to ions can often be more versatily and easily realized with acyclic polyethers than with cyclic ones. Complexation and decomplexation proceed, in general, more rapidly than in cyclic systems, with the pseudocavity formed showing a better conformational flexibility.

Like crown ethers, these systems complex alkaline and alkaline earth cations, but opening of the ring is accompanied by a significant decrease of the complexation constant. For example, the complexation of K^+ decreases by four orders of magnitude on passing from [18]crown-6 *1a* to pentaethyleneglycol dimethyl ether *6* (n = 5). This phenomenon, the so-called "macrocyclic effect" [5], was formerly explained in terms of conformational entropy factors. Later studies [136–138] showed that thermodynamic parameters, in conjunction with solvation effects on the ligand, often play a more important role than conformational entropy factors.

With sufficiently long polyoxyethylene chains containing terminal electron-releasing substituents, particularly stable complexes can be formed with alkaline or alkaline earth cations, which are completely wrapped up by the chains, as shown by X-ray analysis [139].

In line with these results, multidentate podands such as *7* and *8* [140] can operate as catalysts under solid-liquid phase-transfer conditions. A series of solid-liquid PTC nucleophilic displacements was performed in the presence of simple polyglymes *6* with a sufficiently long chain (n = 7–9) [141, 142].

The reactivity of podands increases with increasing molecular weight [141, 143–145]; however, the catalytic activity does not appear to be generalizable for any kind of anion-promoted reaction [146]. Finally, even simple poly(ethylene glycol)s (PEG)

CH₃O(—CH₂CH₂—O)ₙ—CH₃
6

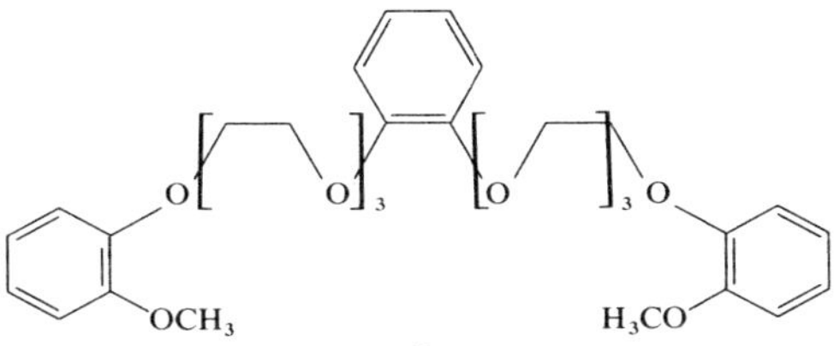

7

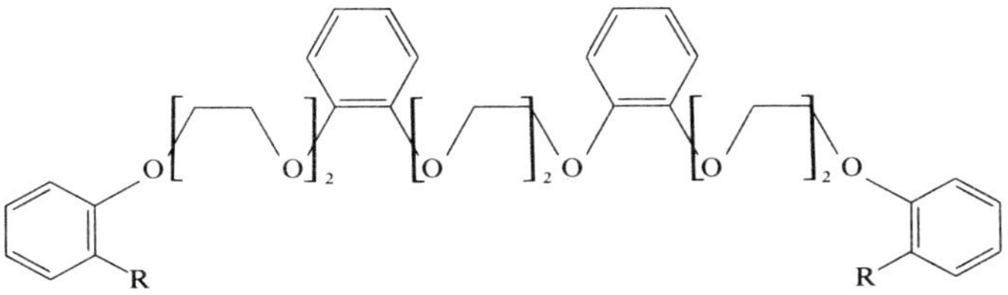

8, R=OCH₃, NO₂

operate, under particular conditions, as PTC catalysts [147-149]: in particular, PEG-1000 can be used instead of [18]crown-6 *1a* in reactions of aryldiazonium salts under PTC conditions, e.g. in the conversion of aryldiazonium tetrafluoroborates into aryl halides and asymmetric biaryls [150].

Neutral branched compounds having many "tentacles" with various electron-releasing atoms (polypodands), e.g. *9*, can complex inorganic cations as in a capsule, thus transferring salts into the organic phase and operating as PTC agents [136-138, 151-153]. Similar properties are shown by cycloveratrylene derivative *10* [154, 155]. Triazine *11* and pentaerythritol *12* derivatives too show the same activity [156]. Finally, by combining the criterion which led to polypodands with the concept of terminal group, ligands *13* and *14* have been synthesized. These can be considered as "open-chain cryptands". In compounds *9* and *11* the catalytic activity is greatly reduced by decreasing the number of electron-releasing groups, reducing both the length

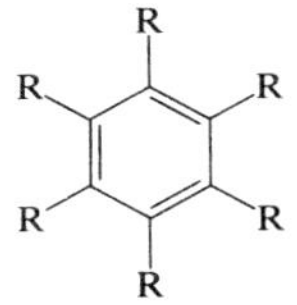

9, R=CH₂S(CH₂CH₂O)₃C₈H₁₇-*n*

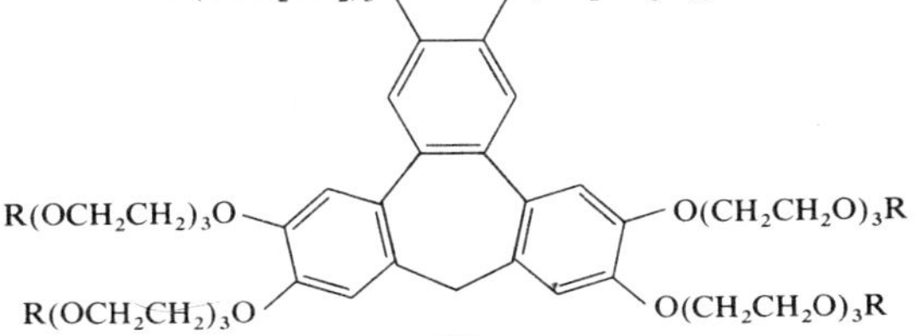

10

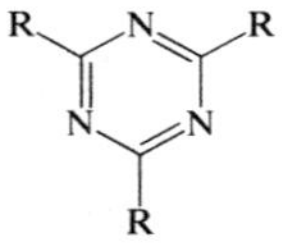

11, R = N[(CH$_2$CH$_2$O)$_4$C$_4$H$_9$-n]$_2$

n-H$_{21}$C$_{10}$OCH$_2$C(CH$_2$OCH$_2$CH$_2$OCH$_2$CH$_2$OCH$_2$CH$_2$OH)$_3$
12

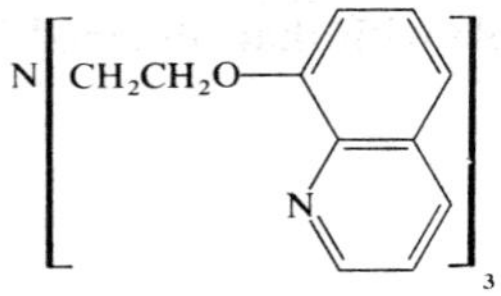

13

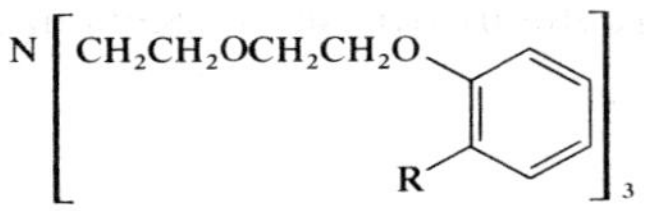

14, R = H, OCH$_3$, NO$_2$

of the "tentacles" and their number. Linear polyethers and polypodands have been mainly used in solid-liquid PTC, but rarely in classical liquid-liquid PTC. It should, however, be pointed out that, even if some of them show a catalytic activity comparable to that of lipophilic macrocyclic ethers, the high molecular weight can constitute a considerable limitation to their use, even with a very low catalyst-to-substrate ratio.

5.4 Other Catalytic Systems

N-alkyl phosphoramides with a long alkyl chain 15 have been successfully used as PTC catalysts. Peculiar to such systems is their versatility in the complexation of different cations; this is in contrast to the selectivity shown by macrocyclic polyethers [157–159].

Phosphorus and sulfur oxides connected by a methylene group, e.g. 16, have analogously been employed as PTC catalysts, although their use seems to be limited to some specific reactions, such as carbanion alkylation [160].

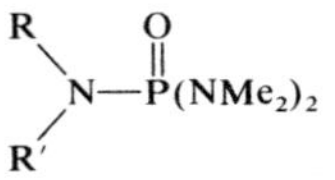

15

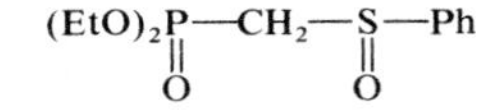

16

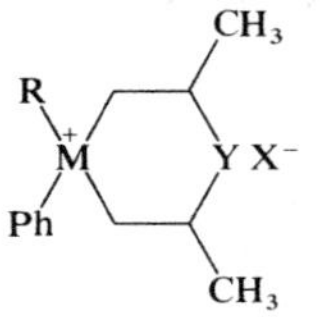

17, M = P, As; Y = O, NR, S

Cyclic phosphonium and arsonium salts 17 have been used as alternatives to analogous open chain systems [161].

Polyamines, such as N,N,N',N'-tetramethylenediamine 18, or more sophisticated amines 19, show a noticeable complexation activity with respect to cations, effectively competing with classical catalysts in PTC reactions. Their use, however, appears to be limited to solid-liquid phase-transfer [122, 157].

Me$_2$N–CH$_2$CH$_2$–NMe$_2$
18

Me$_2$N(CH$_2$)$_n$–N(Me)–(CH$_2$)$_n$NMe$_2$
19, n = 2, 6

n-C$_{12}$H$_{25}$N(O)Me$_2$
20

Catalytic activity probably depends on the formation of quaternary ammonium salts in the reaction mixture [162]. Similarly, the claimed catalytic activity of N-oxides of lipophilic tertiary amines [163], such as *20*, depends on their conversion into a quaternary ammonium salt [164]. 2-Alkylaminopyridinium salts have also been used as catalysts in base-promoted reactions [165].

Even organic derivatives of Ge, Sn, Pb, and in particular $(C_6H_{11})_3SnBr$ and Bu_2SnCl_2, catalyze the hydrolysis of esters in a rate comparable with that obtained with ammonium and phosphonium salts [166]. Finally, polyguanidinium salts, such as *21* and *22*, when complexes with anions [167], are

$$N(CH_2CH_2-X)_3 \qquad \begin{array}{c} CH_2-N(CH_2CH_2-X)_2 \\ | \\ CH_2-N(CH_2CH_2-X)_2 \end{array} \qquad X = C^+(NH_2)_3$$

$$21 \qquad\qquad\qquad 22$$

of considerable interest, due to the possibility of modifying the cationic reactivity.

6 Phase-Transfer in the Presence of Catalysts Bonded to a Polymeric Matrix [168, 169]

6.1 Introduction

Catalysts have been bonded to insoluble polymeric matrices to allow, in principle, noticeable simplifications of PTC: the catalyst is a third insoluble phase which, at the end of the reaction, can be isolated by simple filtration and then recycled, thus avoiding the tedious processes of distillation, chromatographic separation, and so on. This is of potential interest from the industrial point of view, due to the possibility of carrying on both discontinuous processes with a dispersed catalyst and continuous processes with a catalyst on a fixed bed. This technique was introduced by Regen, and called "three-phase-catalysis" [169, 170].

The polymeric matrix normally used is polystyrene, generally cross-linked with 1 to 2% of *p*-divinylbenzene. A number of types of variously functionalized silica have also been used.

The factors affecting the catalytic activity are more numerous than in the case of soluble catalysts, and so their combination to obtain optimum results is much more complex. A fundamental role is played by the chemical stability of the catalyst and the mechanical stability of the matrix, with diffusive factors affecting the reaction rate as well.

6.2 Quaternary Onium Salts. The Influence of Structural Factors and of the Spacer Chain

The direct quaternarization of chloromethylated polystyrene matrices, cross-linked with *p*-divinylbenzene to various degress and/or exhibiting different structures (microporous, macroporous, "popcorn"), is the original and simplest way to obtain PTC catalysts bonded to a polymer [169, 170].

Systems 23, with R = alkyl chain of variable length, catalyzed a number of anion-promoted reactions, i.e. halogen and pseudohalogen exchange, dichlorocarbene generation and addition to double bonds, ether synthesis, dehalogenation of vic-dihalides, oxidation of alcohols to carbonyl compounds with hypochlorite [171]. Even commercial ion-exchange resins 23, with R = CH_3 or CH_2CH_2OH, have been employed in the catalyzed alkylation of carbanions [172].

The importance of structural factors in determining the activity of PTC catalysts bonded to a polymer appears, however, to have been largely underestimated [169]. It is indeed clear that the same factors governing the lipophilicity, and from a certain point of view the anion activation of soluble quaternary salts, must play a fundamental role in the case of immobilized catalysts as well [173, 174]. Short alkyl chains, such as in Me_3N^+, lead to very low catalytic activity, which largely increases in the presence of longer chains, as in the case of Bu_3N^+ and Oct_3N^+. Moreover, ammonium salts were found to be more efficient than the corresponding phosphonium salts [174].

The length of the spacer chain similarly affects the catalytic activity: for example, noticeable reactivity was achieved [175] by bonding the Me_3N^+ group and the polystyrene matrix with long alkyl chains, as in 24.

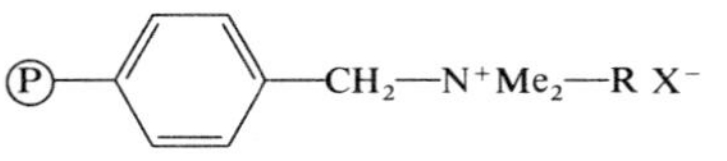

(P)—⟨C₆H₄⟩—CH_2—N^+Me_2—R X⁻

23

(P)—⟨C₆H₄⟩—CH_2—O—$CO(CH_2)_nN^+Me_3$ X⁻

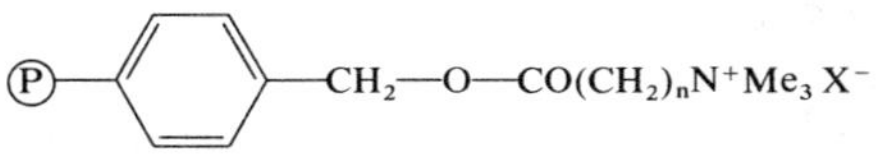

24, n = 5, 11

Even highly lipophilic catalytic centers show low reactivity when directly bonded to the polymeric chain, e.g. 25, n = 0, Q = NBu_3, PBu_3. Reaction rates increase about one order of magnitude when a linear spacer chain of about ten carbon atoms is inserted [173, 176]. By increasing in 25 the chain length up to 20–30 atoms, the rate increases further. This inneage is, however, very small in comparison with the one found when passing from 1 to 10 atoms spacer chain and, under particular conditions, may be equal to zero [174, 176].

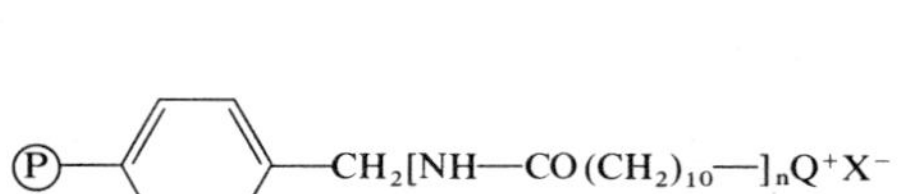

(P)—⟨C₆H₄⟩—$CH_2[NH$—$CO(CH_2)_{10}$—$]_nQ^+X^-$

25, n = 0, 1, 2, 3; Q = NMe_3, NBu_3, PBu_3

Other authors [177] concluded that the length of the spacer chain was irrelevant from the point of view of catalytic activity, but these statements were based on the comparison of reactions carried on under different conditions. Long-spacer chains exhibit a similar positive effect in the classical Merrifield polypeptide synthesis [178].

Table 12. Observed pseudo-first-order rate constants for the reaction of *n*-octyl bromide with KI under PTC conditions in the presence of polymer-supported catalysts.[174]

Catalyst					Catalyst				
Formula	n	Q$^+$	% Cross-linking	$k_{obsd} \times 10^6$ (s^{-1})	Formula	n	Q$^+$	% Cross-linking	$k_{obs} \times 10^6$ (s^{-1})
28	0	$\overset{+}{N}Me_3$	2	0.6	28	0	$\overset{+}{P}Bu_3$	2	118
	0	$\overset{+}{N}Bu_3$	2	157		0	$\overset{+}{P}Bu_3$	4.5	30.6
	0	$\overset{+}{N}Bu_3$	4.5	66		0	$\overset{+}{P}Bu_3$	1	340
	1	$\overset{+}{N}Me_3$	2	20.4		1	$\overset{+}{P}Bu_3$	2	272
	1	$\overset{+}{N}Bu_3$	2	330		2	$\overset{+}{P}Bu_3$	2	308
n-C$_{16}$H$_{33}\overset{+}{N}Bu_3$				950	*n*-C$_{16}$H$_{33}\overset{+}{P}Bu_3$				783

[a] 0.01 Molar equiv; [b] Toluene-H$_2$O, at 90 °C; conditioning time = 15 h; stirring speed 1000 rpm

6.3 Crown Ethers and Cryptands [43)]

In addition to quaternary salts, crown ethers and cryptands have also been analogously bonded to a polymer, both directly, as an integral part of the polymeric chain, and through a spacer chain. Examples of the first group are compounds 26–28.

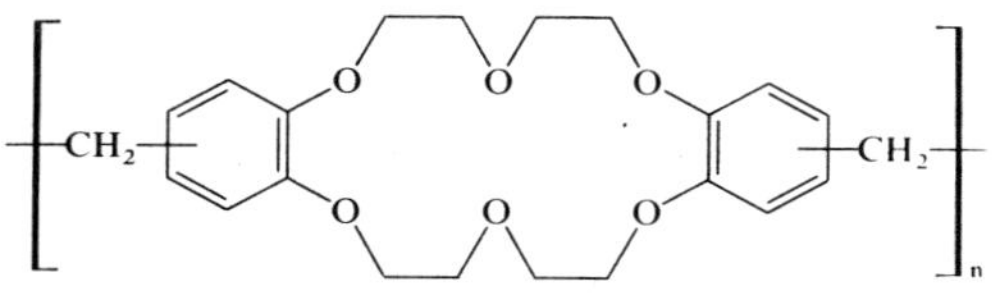

26 [179, 180)

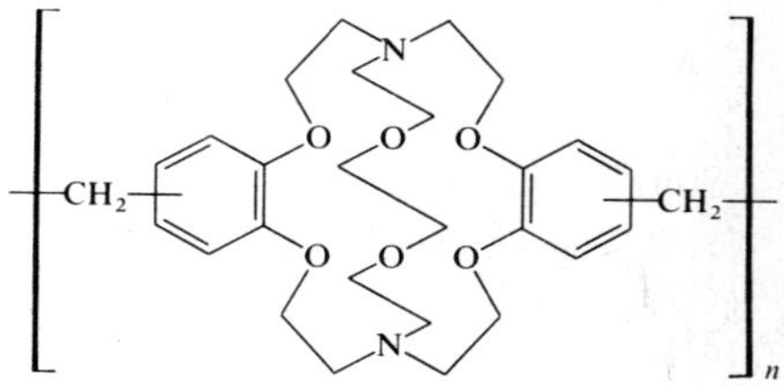

27 [179–181)

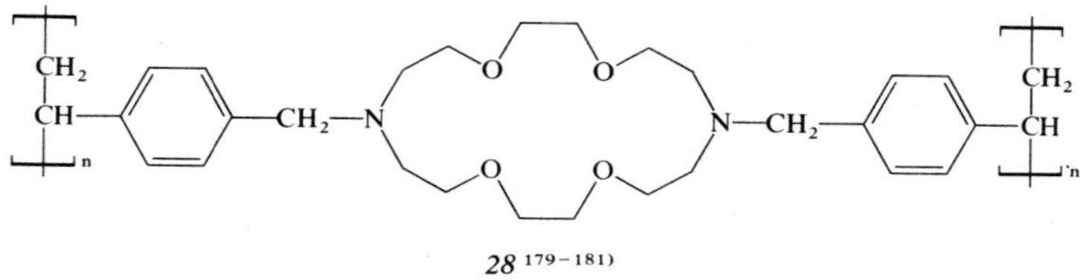

28 [179–181)

Examples of the second group are compounds 29–32.

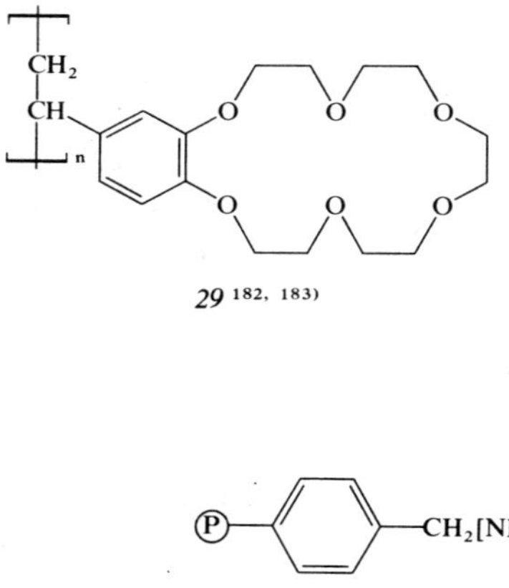

29 [182, 183)

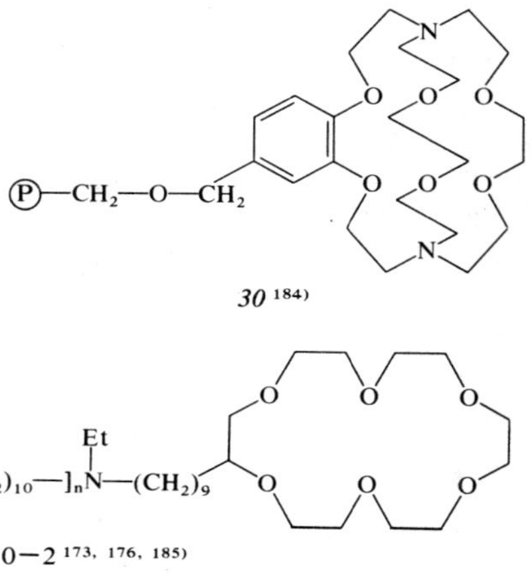

30 [184)

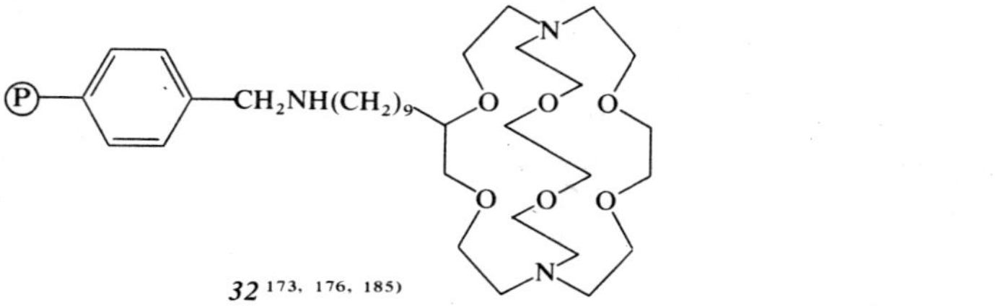

31, n = 0–2 [173, 176, 185)

32 [173, 176, 185)

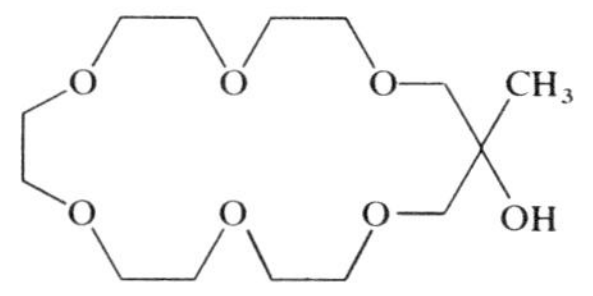

33 [186)]

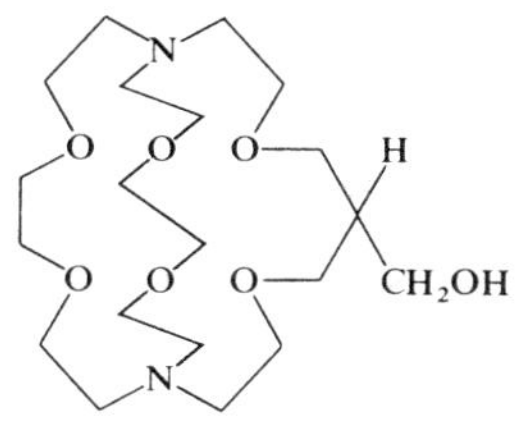

34 [187)]

Suitably functionalized systems, e.g. *1d*, *4d*, *33*, *34* [116, 186, 187)], can be bonded to any polymeric matrix, both directly and through a long interposed chain, or even to hydrocarbon chains to obtain liposoluble homogeneous catalysts.

Macrocyclic and macrobicyclic polyethers bonded to polymers have been used mainly in ionic chromatography or in reverse osmosis membranes [43, 188)]. Recently, they have been more frequently employed in PTC [116, 173, 176, 185–187, 189)].

Catalytic activity apparently follows an order similar to that for soluble catalysts, i.e. cryptands > crown ethers and quaternary salts, under the same reaction conditions [176, 185)]. As expected, crown ethers differ in activity from quaternary salts, as a function of the nature of the anion. The polymeric matrices as such, or rather the linear spacer chains, ensure sufficient lipophilicity.

6.4 Cosolvents Bonded to Polymeric Matrices

Bonding poly(ethylene glycol)monomethyl ethers to a *p*-divinylbenzene cross-linked polystyrene afforded systems 35 and 36 acting as cosolvents, thus influencing the polarity and the freedom of motion of the microenviroments around the reacting species [169, 190–192)]. Interesting results were obtained using radical traps able to distribute themselves between the phases in various ways [193)]. Other types of insoluble cosolvents have been obtained by bonding HMPT [158, 194)] or dimethyl-amino-N-oxides [195)] to a polymeric support (*37*, *38*). All these catalysts can promote reactions similar to those conducted with classical PTC polymer-bonded catalysts. They also work very well under solid-liquid PTC conditions [196, 197)], their reactivity increasing with rising chain length [196)].

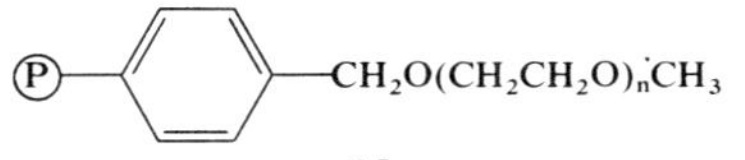

35

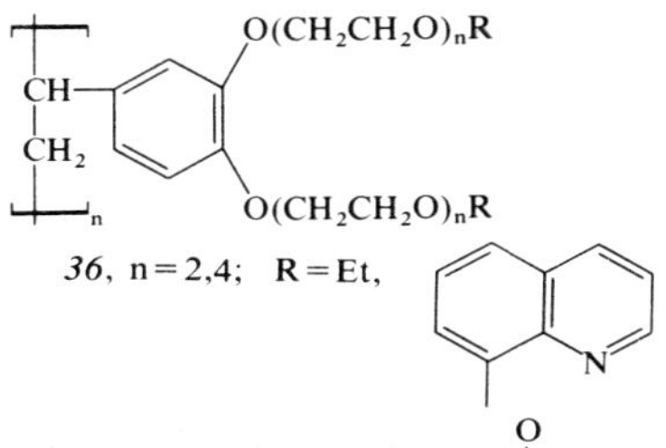

36, n=2,4; R=Et,

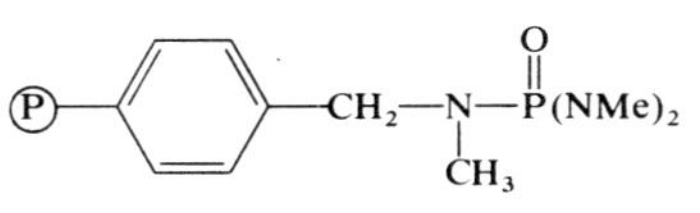

37

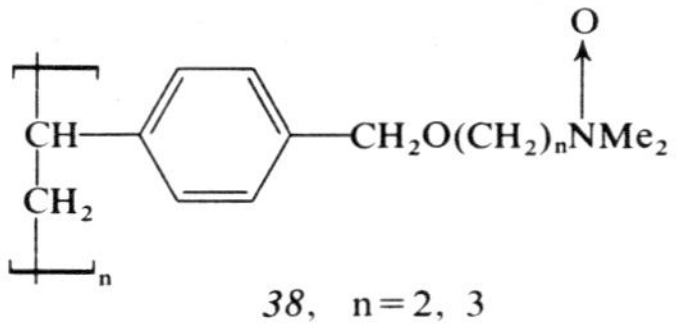

38, n=2, 3

6.5 Swelling of the Resin and Diffusive Factors

Catalysts bonded to a polymer require a preliminary conditioning (3–15 h); in this way, the resin swells by adsorbing solvent to allow the approach of both substrate and reagent to the catalytic center. Swelling depends on the nature of the water-immiscible organic phase, increasing with the polarity of the latter. In relatively polar solvents, maximum swelling is obtained when the groups are directly bonded to the polymeric matrix [174].

The swelling obviously decreases as cross-linking increases. Optimum properties are shown by 1–2% cross-linked polystyrene; swelling, and hence reactivity, are strongly reduced by 4.5% cross-linking. In any case, catalysts bonded to polymers are less reactive than unbonded ones. From this point of view, the influence of diffusive factors is clearly shown by the fact that the progressive increases in stirring rates are accompanied by a contemporary rise in the reaction rates of at least up to 800–900 rpm before reaching a plateau [174]. In liquid-liquid PTC, the catalysts require only 250–300 rpm to attain the plateau. Turbulent stirring sensibly reduces the influence of diffusion, and the plateau is reached at around 600 rpm [198].

According to a scheme proposed by Regen [169], there are microscopic aqueous and organic phases within the polymeric catalysts, whose composition is similar to that of the phases external to the polymer. The polymer can thus generate a large area of separation between the two immiscible phases to increase the interfacial aqueous/organic separation.

This is supported by the fact that, when the polymeric catalyst can settle itself on a large surface of separation between the two liquid phases, the reaction substantially proceeds at the same rate with or without stirring [199]. When a larger amount of catalyst is used with respect to the separation surface, stirring is necessary to increase the contact between the resin and reagents. This behaviour which, in principle, avoids the mechanical grinding of the resins by the stirrer, is likely limited to highly polar and hydrophilic catalysts of relatively low reactivity. It is not found in the case of the best catalysts with a long spacer chain [198].

Friedel-Crafts reactions of α,ω-dihaloalkanes with linear polystyrene afforded a new type of polymer which, when converted into a polymeric quaternary salt, showed high swelling properties and reactivity [200].

6.6 Reaction Mechanism

Several experiments support the hypothesis that the mechanism of PTC reactions in the presence of polymer bonded catalysts is analogous to that of the classic phase-transfer catalysis [174]. It can be described by Eq. (26).

$$
\begin{array}{ccccc}
\text{\textcircled{P}}-Q^+Y^- + RX & \longrightarrow & \text{\textcircled{P}}-Q^+X^- + RY & & \text{(organic phase)} \\
\updownarrow & & \updownarrow & & \\
Y^- & + & M^+ + X^- & & \text{(aqueous phase)}
\end{array}
\qquad (26)
$$

According to this scheme, the reaction should occur in the organic layer which surrounds the catalytic center. The anions are exchanged at the water-organic solvent interface while the M^+ inorganic cation and the Q^+ quaternary cation bonded to the polymer are the corresponding ions in the aqueous and organic phases.

Reaction rates depend on the amount of anion bonded to the resin and are not affected by the amount of anion dissolved in the aqueous phase. Reactions follow pseudo-first-order kinetics (Eq. (27)), and the observed rate constants are linearly related to the equivalents of quaternary groups in the polymer [169, 174, 199].

$$\text{rate} = k_{\text{obs}}[\text{substrate}] \,. \tag{27}$$

Alkylation of phenols and naphthols affords O-alkylation in nearly quantitative yields, which is consistent with classical PTC data [201]. The reaction occurs in the apolar organic phase, thus under Kornblum's selective conditions for O-alkylation [202]. O-alkylation increases with rising length of the spacer chain [174, 175], in agreement with the progressive decrease of polarity around the catalytic center.

Finally, the hydration numbers of the anions bonded to the polymeric cations determined after equilibration with the aqueous and organic phases under reaction conditions, are very similar to those found for lipophilic quaternary salts, complexed crown ethers, and cryptates. This implies that the organic phase permanently surrounds the catalytic centers bonded to the polymer [174].

6.7 PTC Catalysts Bonded to Silica

Quaternary ammonium and phosphonium salts bonded both directly and through spacer chains to silicic matrices operate as PTC catalysts in analogy to quaternary salts bonded to polymers. Many types of silica have been used [203-206]. In the case of catalysts bonded to silica, adsorption phenomena significantly affect the reaction rate. Experiments based on C-/O-alkylation [205, 206] show that reactions occur in a strongly influence of the spacer chain on the reaction rate in this case is absolutely different from that in the case of polystyrene-bonded catalysts. The highest rates are obtained when the catalytic centers are directly bonded to silica, and the observed polar medium, i.e. the functionalized silica and the adsorbed water. Effectively, the rates rapidly diminish with increasing spacer chain length. This change of k_{obs} strictly follows the variation of adsorption constants. With very long chains, k_{obs} again increases, reaching a behaviour similar to that of polystyrene matrices, where the reaction occurs in an apolar organic environment of the catalytic center [205]. In this case, it has been shown that the reaction is not controlled by diffusion. Moreover, the longer the spacer chain, the higher the ratio between the reactivity in the absence of stirring and the optimum rate determined with vigorous mechanical stirring. The rather high ratio obtained in the case of long alkyl chains (0.7) has been explained by assuming that the alkyl chain transports anions from the aqueous to the organic phase, both adsorbed in the third solid phase [206].

6.8 Uses and Chemical Stability

PTC catalysts bonded to a polymeric insoluble matrix have been used for a number of significant organic reactions, under both liquid-liquid and solid-liquid phase-transfer conditions. The principal reasons for bonding PTC catalysts to polymers are both the simplification of the preparative process by easy filtration of the insoluble catalyst and, mainly; the possibility of indefinite recycling of the catalyst.

To achieve these aims: i) minimization of diffusive effects to obtain a reactivity comparable to that of soluble catalysts, ii) high chemical stability of the catalysts bonded to the polymer and iii) high mechanical stability of the polymer matrix must be ensured.

The first goal was partially achieved by the use of long spacer chains [174] or polymers with high swelling capability [200], and/or by using the turbulent stirring technique.

Chemical stability is a factor which should be carefully considered. In the case of quaternary salts, as such or bonded to a polymer, the catalyst can be destroyed both by thermal effects and chemical reactions in the presence of active nucleophiles [207]. This gives rise to some doubts regarding the possibility of the general employment of such catalysts, especially in the industrial field. The situation is completely different in the case of more sophisticated and relatively more expensive catalysts, i.e. crown ethers and cryptands. The latter systems are much more stable than quaternary salts. In principle, polymer bonding allows a broader use, due to the possibility of recycling [185].

The third point deals with the mechanical and chemical stability of the polymeric matrix. Silica-bonded catalysts decompose at high pH [203-205]. Polystyrene matrices show increased swelling properties as cross-linking values decrease; the mechanical stability decreases as well. An optimum compromise between these opposing factors has not yet been realized.

A further factor determining the practical use of such systems is the catalyst-to-substrate ratio. The less polymer used by weight with respect to the substrate and solvent, the easier the reaction proceeds. Large amounts of polymer, when swelled by the solvent, can practically stop the stirring and make large-scale preparation unpractical. The use of small amounts of polymer requires a high catalytic activity, so that one can work with catalyst-to-substrate molar ratios of 1–5% [174]. For this purpose, a relatively high number of catalytic centers along the polymeric chain is needed. This factor has also been largely underestimated for too long; in many reactions reported the quaternary salt bonded to the polymer has been used in nearly stoichiometric amounts, or plainly in excess with respect to the substrate [169,170]. At least in the case of groups directly bonded to the polymer, a high functionalisation value may affect the reactivity which was reported to decrease in the case of a high frequency of the catalytic centers [169,170]. A much more detailed study on the matter is thus of great importance [198].

7 Applications of Phase-Transfer Catalysis

7.1 Introduction

Reviews including full experimental details on practical applications of PTC are included in the books by Weber and Gokel [35], Starks and Liotta [36] and Dehmlow and Dehmlow [37]. A comprehensive register of syntheses by PTC has been published by Fluka Company [32]. Here only few significant examples of PTC applications are schematically reported.

7.2 Displacement Reactions

These reactions, which are largely promoted by anions, are one of the main fields of practical PTC applications. This is one of the sectors in which PTC largely takes the place of classical reactions carried out under homogeneous conditions both in polar protic and in dipolar aprotic solvents.

In the latter case, the advantage of the acceleration of reaction rates is noticeably limited by the high cost of solvents, their difficult elimination and often their toxicity.

Reactions can be carried out under both liquid-liquid and solid-liquid PTC conditions. As the substrates are often liquid under reaction conditions, the use of solvents may be avoided. This is particularly exploited in industry, undoubtedly due to the advantages of saving solvent, shortening reaction times, reducing volumes and simplifying work-up and purification of the products.

Nucleophilic anions, i.e. halides, pseudohalides, alkoxides, phenoxides, and thiophenoxides, are particularly suitable for these reactions. Even anions of lower reactivity in nucleophilic displacements, i.e. carboxylates, nitrates, nitrites and hydroperoxides, find practical application under PTC conditions. Reactions are rigorously S_N2 in mechanism: primary substrates are thus most suitable, since secondary substrates afford elimination products in high yields, especially when reacted at high temperatures, and tertiary substrates only give rise to elimination. This behaviour is consistent with the low polarity of the organic phase, preventing unimolecular mechanisms and favouring elimination over substitution when the reaction center is not a primary carbon atom.

The bimolecular mechanism affects the stereochemistry of substitutions on a secondary carbon. Very high or quantitative inversions were obtained in irreversibile reactions, such as the displacement of the mesyl group by CN^- [11], thiophenoxide [208] and phthalimidate [209], starting from optically active 2-octyl methanesulfonate. Similar reactions with halide ions [210] also proceed with inversion accompanied, however, by racemization. The latter increases with increasing halogen-halogen exchange rate at the chiral center, 5, 14 and 90% with Cl^-, Br^- and I^-, respectively, being obtained. Inversion was found to be predominant with hydroperoxide ion in displacement reactions [211–213]. A series of significant examples of nucleophilic aliphatic displacements is reported in Table 13.

In comparison with the large number of nucleophilic aliphatic substitutions, only few examples of nucleophilic aromatic substitutions carried out under PTC conditions have been published so far.

Halogen atoms and nitro groups have been substituted on aromatic [215–222] and heteroaromatic systems [223–225] by various nucleophiles (halogens, pseudo-halogens, alkoxides, thiophenoxides, carbanions, etc.) when activated by electron-withdrawing groups as well as by chromium carbonyl complexes [226].

The order of nucleofugacity was found to be $F > Cl > Br > I$ [222]. Noteworthy is the introduction of a fluorine atom into the aromatic ring with KF in solid-liquid

Table 13. Nucleophilic displacement reactions [214]

Alkylating agent	Nucleophile	Product	Catalyst and reaction medium[a]	Yield %
RHal	ArO⁻	ROAr	Q^+X^-/LL	70–100
Me$_2$SO$_4$	ROH	ROMe	Q^+X^-/NaOH/LL	73–91
RX (X = Hal, MeSO$_3$)	CN⁻	RCN	Q^+X^- or CE or CR/LL	85–98
RX (X = Hal, MeSO$_3$)	Hal⁻	RHal	Q^+X^- or CE or CR/LL	77–100
RHal	S⁼	R$_2$S	Q^+X^-/LL	81–94
RX (X = Hal, MeSO$_3$)	K-Phthalimide	phthalimide-NR	Q^+X^-/SL	85–99
RHal or R$_2$SO$_4$	indole (NH)	N-alkyl indole	Q^+X^-/NaOH/LL	78–98
RHal	N$_3^-$	RN$_3$	Q^+X^-/LL	75–80
RBr	KO$_2$	ROOR [+ ROH + alkenes]	CE/SL	42–77

[a] Q^+X^- = quaternary onium salt; CE = crown ether; CR = cryptand; LL = liquid-liquid PTC (aqueous-organic two-phase system); SL = solid-liquid PTC

PTC [104, 219] and of a methoxy group with KOH in methanol [220, 222, 227].

The latter reagent allowed the substitution on some poorly activated substrates, such as *o*- and *m*-dichlorobenzene, to give, although in moderate yields, *o*- and *m*-chloroanisole, respectively. This shows that the reaction involves direct attack of the nucleophile at the reaction center, not a benzyne-type mechanism [227].

These intermediates cannot be detected in liquid-liquid PTC where the water associated to the anion in the organic phase assists the cleavage of the bond between the leaving-group and the aromatic carbon atom [222].

7.3 Alkylations

Alkylations occur under PTC conditions in the presence of bases (generally NaOH and KOH used both in aqueous 30 to 50% solution and in the solid state): these reactions have most extensively been studied in the PTC field. The base generates the carbanion at the interface, according to Makosza's and Dehmlow's data [7, 19, 95–98]. The cation, both quaternary organic and metallic complexed by polyethers, then selectively transports the lipophilic carbanion into the organic phase, yielding the products. One of the principal advantages is the elimination of any side reaction of hydrolysis in the organic phase, since the latter is nearly dry under the reaction conditions; the concentration of OH⁻ in this phase is indeed negligible [95]. Possible hydrolytic processes only occur at the interface.

Particularly interesting is also the technique of extractive alkylation developed by Brändström [9], requiring a stoichiometric amount of a quaternary salt, usually $Bu_4N^+HSO_4^-$. With regard to the catalytic process, this technique shows the possibility of using highly reactive alkylating agents, such as alkyl iodides, although the latter are not practically applied in PTC, due to the striking lipophilicity of the leaving group [11]. Carboxylic acids with pK_a up to 22–25 [95] can by alkylated under these conditions. Examples are alkanenitriles and isonitriles, malonic esters, arylacetic esters and amides, carbonyl compounds, and conjugated acids of aromatic anions such as cyclopentadienyl and fluorenyl.

Table 14. C-Alkylations under PTC conditions [230]

Substrate	Alkylating agent	Product	Catalyst and reaction medium[a]	Yield %
ArCHCN with R (R = H, alkyl, aryl)	R'Hal (R' = alkyl, aryl, CH_2CO_2R'', CH_2CN, etc.)	ArCCN with R and R'	Q^+X^-/NaOH/LL	26–95
$ArCH_2CN$	$Br(CH_2)_nBr$ (n = 3–5)	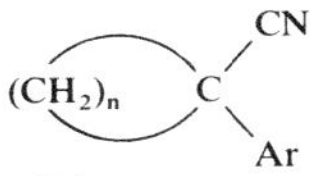 $(CH_2)_n$C(CN)(Ar) cyclic with R'	Q^+X^-/NaOH/LL	26–88
RCHCN $\|$ X (R = Ph, PhS; X = NMe_2, O alkyl)	R'Hal (R' = alkyl)	RCCN with R' and X	Q^+X^-/NaOH/LL	12–82
$(CN)CH_2COOBu^t$	RBr (R = alkyl)	$(CN)CHCOOBu^t$ $\|$ R	Q^+X^-/NaOH/LL	67–87
$CH_2(COOEt)_2$	RHal (R = alkyl)	$RCH(COOEt)_2$	Q^+X^- or CE/K_2CO_3/SL	93–94
PhCHCOR' with R (R = H, alkyl; R' = alkyl, Ph)	R"Hal (R" = alkyl)	PhCCOR' with R and R"	Q^+X^-/NaOH/LL	40–90
RR'CHCHO (R, R' = alkyl)	R"Hal (R" = alkyl)	RR'R"CCHO	Q^+X^-/NaOH/LL	15–85
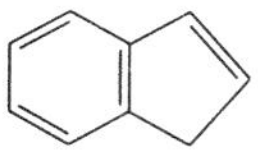 indene	RBr (R = alkyl)	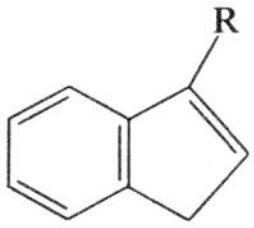 R-substituted indene	Q^+X^-/NaOH/LL	45–73
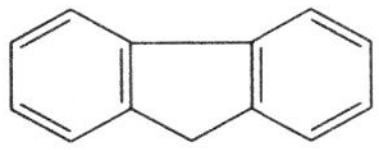 fluorene	RBr (R = alkyl)	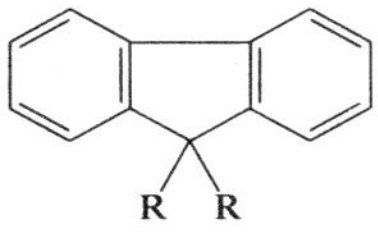 9,9-R,R-fluorene	Q^+X^-/NaOH/LL	80–83

[a] Q^+X^- = quaternary onium salt; CE = crown ether; LL = liquid-liquid PTC (aqueous-organic two-phase system); SL = solid-liquid PTC

Compounds with activated methylene groups usually give mixtures of mono- and bis-alkylation products; when reacted with α,ω-disubstituted alkylating agents, the main product is a cycloalkane.

Aromatic halogeno derivatives activated by electron-withdrawing groups easily react with carbanions under the described conditions.

Substrates which are precursors of ambidentate anions, i.e. carbonyl compounds or phenols, can undergo both C- and O-alkylation [228]. Under PTC conditions, the C-/O-alkylation ratio depends on a number of factors, such as ion pair association, anion conformation, solvation at oxygen or carbon, and nature of the alkylating agent [229].

The situation is much simpler in the case of the alkylation of phenols and naphthols where O-alkylation is classically favoured by poor solvation of the more electronegative oxygen atom, for example in dipolar aprotic solvents [202]. In line with this, PTC reactions yield nearly exclusively O-alkylation products [174,201]. Significant examples of alkylation reactions are listed in Table 14.

7.4 Eliminations

Under PTC conditions, secondary substrates, and in particular cyclohexyl derivatives, yield elimination products in appreciable to quantitative yield. Under similar conditions, tertiary substrates afford only alkenes [231]. vic-Dibromoalkane conversion to olefins catalyzed by I^- in the presence of thiosulfate, the latter continuously reducing the produced I_2, is a further example of the versatility of PTC for such reactions [232]. F^- ion has been used as a base, particularly under solid-liquid PTC conditions [233,234]. Examples of elimination reactions are reported in Table 15.

Table 15. Elimination reactions [231]

Substrate	Product	Catalyst and reaction medium[a]	Yield %
RCHCHR' | | Br Br (R, R' = H, alkyl, aryl)	RCH=CHR'	Q^+X^-/$Na_2S_2O_3$, I^-_{cat}/LL	84—95
RCH_2CH_2Hal	$RCH=CH_2$	CE/Bu^{t}OK/SL	73—97
$PhCH_2CH_2Br$	$PhCH=CH_2$	Q^+X^-/NaOH/LL	100
RCHCHR' | | Hal Hal (R, R' = H, alkyl)	RC≡CR'	CE/Bu^{t}OK/SL	66—98
ArCH=CXR (R = H, Me; X = Cl, Br)	ArC≡CR	CE/KF/SL	15—80

[a] Q^+X^- = quaternary onium salt; CE = crown ether; LL = liquid-liquid PTC (aqueous-organic two-phase system); SL = solid-liquid PTC

7.5 α-Eliminations: Carbenes and Analogous Compounds

Among the oldest and more spectacular applications of PTC is the generation of carbenes (mostly dihalocarbenes) from haloforms, aqueous concentrated alkaline hydroxide and catalytic amounts of a quaternary salt (generally TEBA). The carbenes formed undergo no hydrolytic side reactions [95] and can therefore react under optimum conditions with the substrate, via addition or insertion. Reaction conditions are much simpler and yields are always noticeably higher than in the classical synthesis run in homogeneous medium. Moreover, it is possible to generate new carbenes that cannot otherwise be prepared. Carbenes can also be formed from $Cl_3C-CO_2^-$

Table 16. Addition of carbenes to alkenes, cycloalkenes, allenes, and alken-ynes [236]

Substrate	Source of carbene	Product	Catalyst and reaction medium[a]	Yield %
$PhCH=CH_2$	$CHCl_3$	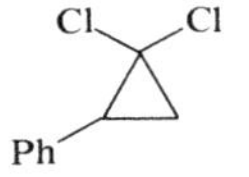 Cl,Cl-cyclopropane with Ph	Q^+X^-/NaOH/LL	80–95
$(CH_2)_n$ (n = 3, 5, 6, 8)	CCl_3COONa	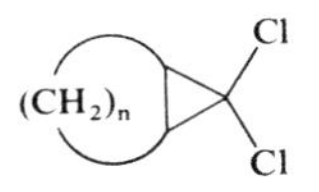 bicyclic Cl,Cl-cyclopropane with $(CH_2)_n$	Q^+X^-/SL	80–88
$RR'C=C=CRR'$ (R, R' = H, Me)	$CHCl_3$	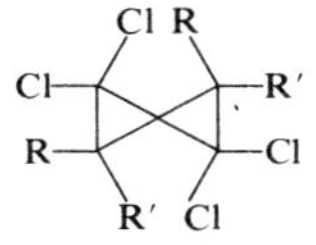 bis-cyclopropane Cl,R / R,Cl / R',Cl	Q^+X^-/NaOH/LL	34–90
$CH_2=C(Me)C\equiv$ $\equiv CC(Me)=CH_2$	$CHCl_3$	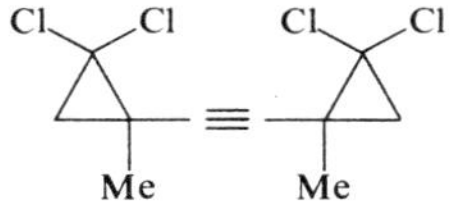 two Cl,Cl-cyclopropanes linked by $\equiv$, Me substituents	Q^+X^-/NaOH/LL	62
$RCH=CHR'$ (R, R' = H, alkyl)	$CHClBr_2$	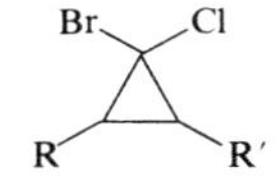 Br,Cl-cyclopropane with R, R'	CE/NaOH/LL	44–62
$RCH=CHMe$ (R = alkyl, Ph)	$CHFBr_2$	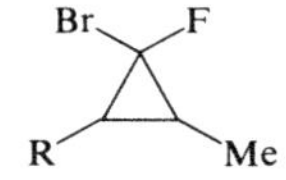 Br,F-cyclopropane with R, Me	Q^+X^-/NaOH/LL	69–88
$RCH=CH_2$ (R = alkyl)	$CHBr_3$	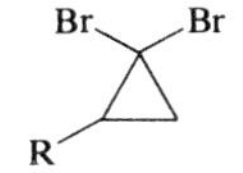 Br,Br-cyclopropane with R	Q^+X^-/NaOH/LL	78
$ArRC=CH_2$ (R = H, alkyl)	CHI_3	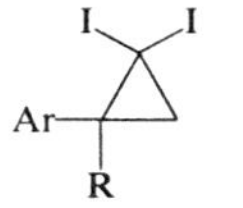 I,I-cyclopropane with Ar, R	Q^+X^-/NaOH/LL	20–59
$RCH=CHR'$ (R, R' = H, alkyl, Ph)	$PhSCHClX$ (X = H, Cl)	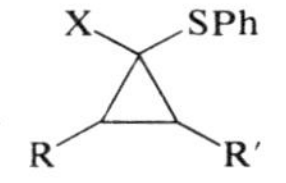 X,SPh-cyclopropane with R, R'	Q^+X^-/NaOH/LL	63–79

[a] Q^+X^- = quaternary onium salt; CE = crown ether; LL = liquid-liquid PTC (aqueous-organic two-phase system); SL = solid-liquid PTC.

which, when transported into the organic phase, decomposes to CO_2 and $:CCl_2$ [100, 101].

Crown ethers, cryptands, quaternary salts bonded to a polymer (Sects. 5.1, 5.2, 6) and finally free amines have been successfully used in carbene generation in two-phase systems. In the latter case, the mechanism is rather complex, and different interpretations have been advanced. However, the last step of the generation of the dihalocarbene probably does not differ substantially from the reaction of $CHCl_3$, NaOH and quaternary salt [235]. In Table 16 are compiled some examples of addition and insertion of carbenes.

Dihalocarbenes produced under PTC conditions have been advantageously used in the classical isonitrile synthesis from primary amines, again giving higher yields than the classical procedure. Hydrazine affords diazomethane, amides are dehydrated to nitriles and sulfoxides are reduced to sulfides in high yields. Finally, alcohols are converted into halides (Table 17) [236].

Table 17. Other reactions of carbenes [236]

Substrate	Source of carbene	Product	Catalyst and reaction medium[a]	Yield %
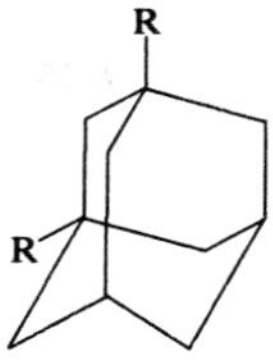 Adamantane (R = H, Me)	$CHCl_3$	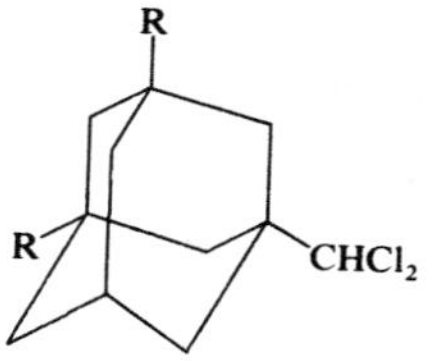 Adamantyl-$CHCl_2$	Q^+X^-/NaOH/LL	91–100
R—NH_2 (R = alkyl, aryl)	$CHCl_3$	RNC	Q^+X^-/NaOH/LL	40–60
RNSO (R = cyclohexyl, aryl)	$CHCl_3$	RNC	CE/KOH/LL	75–96
NH_2NH_2	$CHCl_3$	CH_2N_2	CE/NaOH/LL	48
$RCONH_2$ (R = alkyl, PhCH=CH)	$CHCl_3$	RCN	Q^+X^-/NaOH/LL	40–95
ROH (R = alkyl, adamantyl)	$CHCl_3$	RCl	Q^+X^-/NaOH/LL	90–94

[a] Q^+X^- = quaternary onium salt; CE = crown ether; LL = liquid-liquid PTC (aqueous-organic two-phase system)

7.6 Other Base-Promoted Reactions: Isomerizations, H/D Exchanges, Nucleophilic Additions, Condensations

Bases under PTC conditions promote both the conversion of acetylenes into allenes and the isomerisation of double bonds. H/D exchange can be easily obtained on every substrate capable of generating an intermediate carbanion. Compounds with

Table 18. Isomerizations, H/D exchanges, additions to unsaturated systems, reactions of carbonyl compounds [237]

Substrate	Product	Catalyst and reaction medium[a]	Yield %
PhRCHC≡CPh	PhRC=C=CHPh	Q^+X^-/KOH/SL	100
n-$C_6H_{13}$$COCH_3$	n-$C_5H_{11}$$CD_2$$COCD_3$	Q^+X^-/NaOD, D_2O/LL	99
fluorene	9,9-dideuterofluorene	Q^+X^-/NaOD, D_2O/LL	98
2-ethylthiazole	2-deutero-5-ethylthiazole	Q^+X^-/NaOD, D_2O/LL	90
PhC≡CH	$(PhCH=CH)_2S$	CE/Na_2S/LL	40
PhCHO	Ph-glycidate (Ph/COOEt epoxide)	CE/K_2CO_3, $ClCH_2COOEt$/SL	72
RCOR' (R, R' = H, alkyl, Ph)	R,R' epoxide with CN	Q^+X^-/NaOH, $ClCH_2CN$/LL	10–80
$PhSO_2CH=CH_2$	cyclopropane with $PhSO_2$, Me, CN	Q^+X^-, NaOH, MeCHClCN/LL	7–75
$R_2C=CHCHO$ (R = H, alkyl, Ph)	$R_2C=CHCH=C(Z)COOR'$ (Z = MeCO, EtOCO, CN)	Q^+X^-/ZCH_2COOR', K_2CO_3/SL	14–91
PhCHO	PhCOCHOHPh	Q^+CN^-/LL	70
ArCHO	ArCHOHCOOH	Q^+X^-/$CHCl_3$, NaOH/LL	75–83
ArCHO	$ArCHNH_2COOH$	Q^+X^-/$CHCl_3$, LiCl, NH_3, KOH/LL	29–81
cyclohexene	1-iodo-2-X-cyclohexane (X = N_3, SCN)	Q^+X^- or CE/I_2, KX/LL	43–50

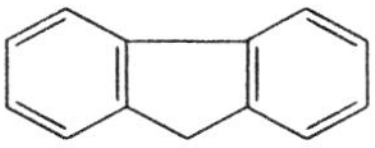

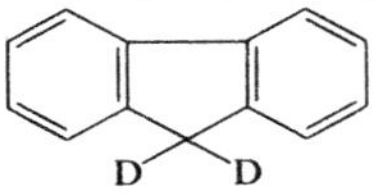

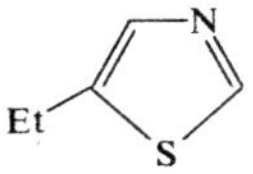

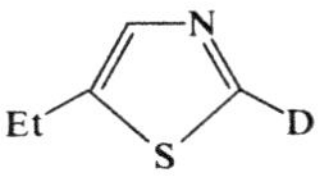

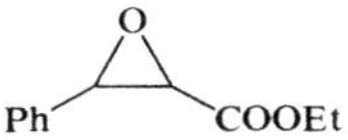

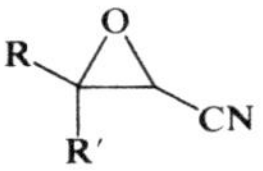

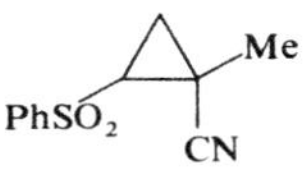

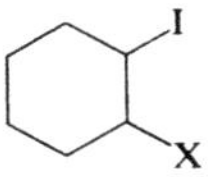

[a] Q^+X^- = quaternary onium salt; CE = crown ether; LL = liquid-liquid PTC (aqueous-organic two-phase system); SL = solid-liquid PTC

carbon-carbon triple bonds add nucleophiles and those with activated carbon-carbon double bonds undergo Michael reactions. Darzens reactions, benzoin and aldol condensations may be carried out similarly (Table 18) [237].

Wittig and Wittig-Horner reactions, classically requiring strictly anhydrous conditions, can be run much more easily in PTC. In these cases, the intermediate operates as a PTC agent, and the presence of catalysts is therefore not necessary. Typical sulfonium and sulfoxonium ylide reactions have been reproduced, yielding oxiranes and cyclopropanes (Table 19) [237].

Carboxylic esters have been hydrolyzed to acids in the presence of quaternary salts, crown ethers and cryptands. Of interest in this regard is the activation of KOH by the [2.2.2] cryptand under solid-liquid PTC conditions [123].

Table 19. Wittig, Wittig-Horner and sulfonium and oxosulfonium ylide reactions [237]

Substrate	Reagent	Product	Catalyst and reaction medium[a]	Yield %
RCHO (R = alkyl, aryl)	R'CH$_2$P$^+$Ph$_3$Hal$^-$ (R' = alkyl, aryl, vinyl)	RCH=CHR'	NaOH/LL	15–94
CH$_2$O	ArCH$_2$P$^+$Ph$_3$Hal$^-$	ArCH=CH$_2$	NaOH/LL	87–98
RCHO (R = PhCH=CH, aryl)	(EtO)$_2$P(O)CH$_2$R' (R' = PhCH=CH, aryl, SR", SOR", SO$_2$R")	RCH=CHR'	Q$^+$X$^-$ or CE/NaOH/LL	12–84
RCOR' (R, R' = alkyl, aryl)	(EtO)$_2$P(O)CH$_2$R" (R" = CN, COOR, aryl)	RR'C=CHR"	Q$^+$X$^-$/NaOH/LL	51–77
PhCOR (R = H, alkyl, aryl)	(O)$_n$S$^+$Me$_3$I$^-$ (n = 0, 1)	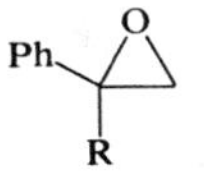 Ph–(epoxide)–R	Q$^+$X$^-$/NaOH/LL	18–92

[a] Q$^+$X$^-$ = quaternary onium salt; CE = crown ether; LL = liquid-liquid PTC (aqueous-organic two-phase system)

7.7 Reactions Promoted by Aqueous Hydrohalic Acids

Lipophilic quaternary salts show high catalytic activity in reactions promoted by aqueous hydrohalic acids. Under liquid-liquid PTC conditions, a number of reactions have been realized: the conversion of alcohols into alkyl halides [238], the cleavage of ethers [239], the addition of hydrohalic acids to alkenes according to Markownikow's rule [240], and acid hydrolysis of esters [241].

The mechanism of the reaction likely requires the formation of an addition complex [242] (Eq. (28)) which, when extracted into the organic phase, reacts with the substrate.

$$Q^+Hal^-_{org} + H^+Hal^-_{aq} \rightleftharpoons Q^+HHal^-_{2\,org} \tag{28}$$

It has recently been found that an appropriately prepared complex of tetraethylammonium bromide with hydrogen bromide reacts with alkynes under anhydrous homogeneous conditions to yield bromoalkenes; the addition product follows Markownikow's rule [243].

7.8 Oxidations

7.8.1 Permanganate

Oxidizing anionic species, transferred into an organic medium by a PTC agent, normally show higher reactivity than under classical conditions.

Well known examples are oxidations with MnO_4^- where the latter is transferred into an apolar organic medium such as benzene, "purple benzene", by complexed macrocyclic polyethers or quaternary cations, both in liquid-liquid [11, 15] and solid-liquid phase-transfer systems [244]. This procedure eliminates the self-catalyzed decomposition of MnO_4^- which occurs with evolution of O_2. Olefins, alcohols, alkylated arenes, etc. are thus oxidized to carboxylic acids, ketones, aldehydes, or glycols.

Reactions are generally exothermal. The oxidation is extremely sensitive to pH in the aqueous phase. Thus, olefins are converted into glycols instead of carboxylic acids when the aqueous phase is a strongly basic $KMnO_4$ solution [245, 246]. This different behaviour has not as yet been explained. It should, however, be remembered that some oxidations, requiring acidic or basic catalysis, proceed with difficulty under PTC conditions [247].

7.8.2 Chromate

Alcohols, particularly benzylic and allylic ones, are oxidized to aldehydes and ketones under PTC conditions. Both stoichiometric and catalytic processes have been performed [248, 249]. As in the case of permanganate, the reaction is controlled by pH in the aqueous phase. The selective oxidation of primary alcohols to aldehydes is an example. This procedure avoids the side formation of carboxylic acids and esters [249]. Under homogeneous conditions, *bis*-tetrabutylammonium dichromate quantitatively oxidizes activated primary halides and secondary alkyl bromides to the corresponding carbonyl compounds [250].

7.8.3 Hypochlorite

This very inexpensive oxidant has been successfully used in the conversion of benzylic and benzhydrylic alcohols to aldehydes and ketones. α-Monosubstituted and α,α'-disubstituted primary amines yield nitriles and aldehydes, and ketones, respectively [251]. Recently, the synthesis of aliphatic isocyanates via a two-phase Hofmann reaction of amides has been performed [252].

Table 20. Oxidations with permanganate, chromate and hypochlorite [253]

Substrate	Product	Catalyst and reaction medium[a]	Yield %
$RCH=CH_2$	RCOOH	$Q^+X^-/KMnO_4/LL$	81–91
cyclohexene	$HOOC(CH_2)_4COOH$	$CE/KMnO_4/SL$	100
PhCH=CHPh	PhCOOH	$Q^+X^-/KMnO_4/LL$	95
RCH=CHR'	RCHOHCHOHR'	$Q^+X^-/KMnO_4$, NaOH/LL	40–80
RC≡CH	RCOOH	$Q^+X^-/KMnO_4/SL$	61–90
ArMe	ArCOOH	$CE/KMnO_4/SL$	78–100
$PhCH_2CN$	PhCOOH	$Q^+X^-/KMnO_4/LL$	86
RCH_2OH (R = alkyl, aryl)	RCHO	$Q^+X^-/K_2Cr_2O_7, H_2SO_4/LL$	78–98
$ArCH_2OH$	ArCHO	$Q^+X^-/NaOCl/LL$	47–100
$RR'CHNH_2$ (R, R' = alkyl)	RR'CO	$Q^+X^-/NaOCl/LL$	84–98
RCH_2NH_2	RCN	$Q^+X^-/NaOCl/LL$	60–76
PhCOCH=CHAr	PhCO-(epoxide)-Ar	$Q^+X^-/NaOCl/LL$	66–100
$RCONH_2$	RNCO	$Q^+X^-/NaOCl/LL$	16–87

[a] Q^+X^- = quaternary onium salt; CE = crown ether; LL = liquid-liquid PTC (aqueous-organic two-phase system); SL = solid-liquid PTC

7.8.4 Osmium and Ruthenium Tetroxides

A large number of oxidations can be performed with OsO_4 and RuO_4. Although they are very strong oxidants in their higher oxidation state, they are practically not used in stoichiometric amounts because of their high cost and toxicity. The use of catalytic amounts of osmium and ruthenium salts in the presence of an excess of less expensive oxidants, capable of oxidizing continuously both elements to the higher oxidation state, is a well known technique in organic chemistry [254, 255]. Due to the striking solubility of both tetroxides in low-polarity organic solvents, working under PTC conditions has proved very advantageous. Thus in the presence of catalytic amounts of osmium salts, olefins mainly afford aldehydes. Under similar conditions, ruthenium salts furnish carboxylic acids in quantitative yield [256, 257].

Table 21. Other oxidations using osmium and rutenium tetroxides, hydrogen peroxide, peroxides, and oxygen [253]

Substrate	Product	Catalyst and reaction medium[a]	Yield %
$n\text{-}C_6H_{13}CH=CH_2$	$n\text{-}C_6H_{13}CHO$	Q^+X^-/OsO_4 cat., H_5IO_6	73 (RCO_2H 13%)
$n\text{-}C_6H_{13}CH=CH_2$	$n\text{-}C_6H_{13}CO_2H$	Q^+X^-/RuO_4 cat., H_5IO_6	94
 cyclohexene	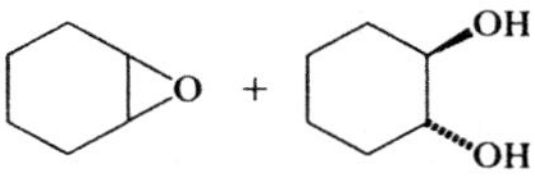 cyclohexene oxide + trans-cyclohexane-diol	Q^+X^-/OsO_4 (or MoO_3, H_2WO_4) cat., H_2O_2/LL	100
$ArCH=CHCAr$ (O)	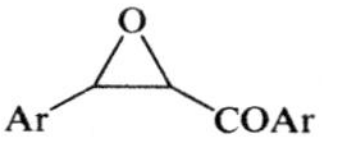 Ar—epoxide—COAr	Q^+X^-/H_2O_2/LL	66–100
$n\text{-}C_6H_{13}CH=CH_2$	$n\text{-}C_6H_{13}CHHal\text{-}CH_2Hal$	$Q^+X^-/HHal$, H_2O_2/LL	56–96
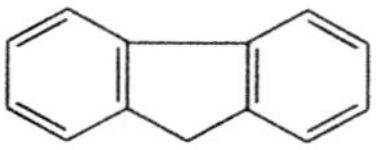 fluorene	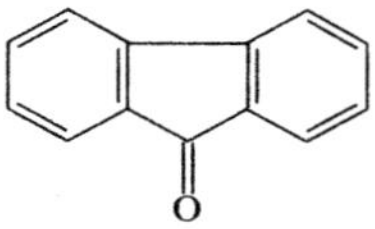 fluorenone	$Q^+X^-/NaOH$, O_2/LL	100
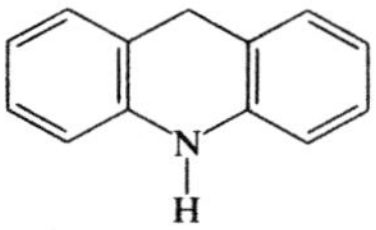 acridine (NH)	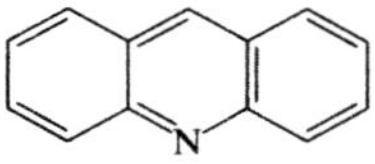 acridone	$Q^+X^-/NaOH$, O_2/LL	97
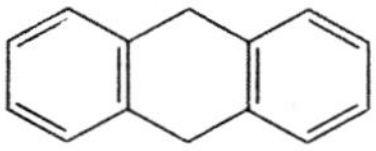 anthracene	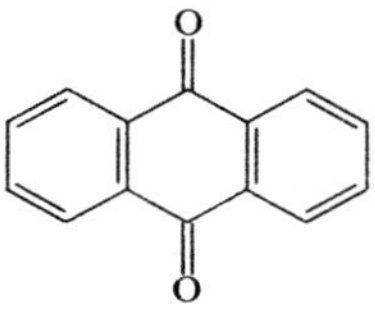 anthraquinone	$Q^+X^-/NaOH$, O_2/LL	100
$PhNH_2$ $RCOCHMe_2$	$PhN=NPh$ $RCOCMe_2OOH$	CE/KOH, O_2/SL $Q^+/X^-/NaOH$, O_2/LL	35–40 15–20

[a] Q^+X^- = quaternary onium salt; CE = crown ether; LL = liquid-liquid PTC (aqueous-organic two-phase system); SL = solid-liquid PTC

7.8.5 Hydrogen Peroxide, Peroxides and Oxygen

Sparse data in the literature indicate that in principle H_2O_2 can be used as an oxidant in PTC. It has, however, been pointed out [242] that it is difficult to hypothesize the extraction of highly hydrophilic HO_2^- by a lipophilic cation, in line with the data obtained for OH^-. Hydrogen peroxide as such can indeed be transferred into the organic phase as a solvate of the anion associated to the catalyst, analogously to water. Complex mixtures of oxidized species have been found in the H_2O_2 oxidation of cyclohexene under PTC conditions in the presence of a series of metallic cocatalysts [258, 259]. vic-Dihalo derivatives have been obtained starting from olefins and hydrohalic acids with H_2O_2 as oxidizing agent, the latter generating in situ Hal_2 [260].

Carbanions generated under PTC conditions have been oxidized with air. Under these conditions, systems such as indene, fluorene, dihydroanthracene, dihydroacrydine, etc. have been oxidized to the corresponding carbonyl compounds or aromatized [123, 261]. Aniline and unsaturated ketones have been oxidized to azobenzene [38] and to oxo-vinyl hydroperoxides [262], respectively. Significant examples of oxidations are reported in Tables 20 and 21.

7.9 Reductions

7.9.1 Complex Hydrides

Because of its significant stability in water, sodium borohydride is the most interesting agent for reductions under PTC conditions.

Usual lipophilic quaternary salts catalyze the reduction of carbonyl compounds to alcohols with $NaBH_4$ in a two-phase aqueous-organic system, the reactions proceeding rather slowly [11]. A substantial improvement is noted using particularly ammonium salts with a hydroxy group β to the cationic center in one of the alkyl chains bonded to nitrogen, typically N,N-dialkylepherinium salts [93, 94]. The effect of the hydroxy group is likely to be due to electrophilic activation, as shown in 39.

This is consistent with data obtained for the corresponding amphetaminium salts of similar structure but with no OH group. In this case, the reaction rate decreases drastically [93, 94].

Crown ethers and cryptands can also be used for such reactions [112, 113, 124, 128]. In this regard is should be mentioned that, in homogeneous organic phase, the complexation of $LiAlH_4$ and $NaBH_4$ by specific cryptands ([2.1.1] and [2.2.1] for Li^+ and Na^+, respectively) inhibits or greatly slows down the reduction of ketones, due to the absence of the electrophilic activation by the inorganic cation [263-266].

An interesting utilization of the facile extraction of BH_4^- into an organic medium by quaternary cations is the in situ generation of diborane by reaction of "acids" (i.e.

BF$_3$, AlCl$_3$, H$_2$SO$_4$, or alkyl halides) with an anhydrous solution of Bu$_4$N$^+$BH$_4^-$ in CH$_2$Cl$_2$ [267]. This reaction has been successfully applied to the hydrogenolysis of halides and sulfonate esters to alkanes and to the anti-Markownikow hydration of olefins under liquid-liquid PTC conditions [268].

7.9.2 Other Reducing Agents

Formamidinesdulfinic acid, H$_2$N—C(=NH)—SO$_2$H, a well known reducing agent synthesized by H$_2$O$_2$ oxidation of thiourea, has been successfully employed under PTC conditions in the reduction of disulfides and sulfylimines to thiols and sulfides, respectively [269].

Nitroarenes are reduced to anilines with Fe$_3$(CO)$_{12}$ [270] and conjugated dienes can be selectively hydrogenated in 1,4-positions with K$_3$[Co(CN)$_5$H] [271].

7.10 Applications of PTC to Transition Metal Chemistry

Compared to the applications in the field of organic chemistry, the use of PTC in organometallic chemistry is much more limited. An increasing number of applications, however, appeared in recent years, and it is clear that even in this field the use of PTC presents extraordinary advantages from the point of view of both synthesis and reactivity.

An excellent review has recently been published on the matter [272]; more detailed information may be obtained there.

Here, we report only some significant examples. The carbonylation of organic halides with carbon monoxide in the presence of transition metals and bases affords carboxylic acids or the corresponding salts or esters according to reaction conditions. By working in a PTC system, salts of carboxylic acids have been obtained in excellent yields in the presence of Pd, Pt or Ni complexes [273-275]. This reaction has been applied to benzylic, vinylic, aromatic and heterocyclic halides. For example, phenylacetic acid has been isolated with a conversion of over 4000 mol per mol of Pd and 10 mol of PPh$_3$ [273] (Eq. (29)).

$$C_6H_5CH_2X + CO + NaOH \xrightarrow[Pd(Ph_3P)_3, Q^+X^-]{toluene/H_2O} C_6H_5CH_2COONa \quad (29)$$

p-Dibromobenzene has been selectively converted into p-bromobenzoic acid in 90–95% yield [273]. In some cases, double carbonylation of benzylic halides has been realized, affording arylpyruvic acids [275,276]. Butanoic acids have been obtained similarly by carboxylation of allylic halides [277].

Aromatic nitriles are classically prepared by reaction of activated aryl halides and CN$^-$ in the presence of complexes such as Ni(Ph$_3$P)$_3$. The same reaction has been realized under PTC conditions in high yields, with a high turnover of the catalyst [278].

Acrylonitriles have been synthesized from vinyl halides and Pd(PPh$_3$)$_3$ complexes [279]. Finally, ferrocene has been easily obtained in the presence of [18]crown-6 under PTC conditions [280].

8 Asymmetric Syntheses

In principle, PTC is an extremely interesting technique for asymmetric syntheses, using chiral catalysts. The crucial factor is indeed the possibility of using catalytic amounts of chiral agents, since the latter are used in stoichiometric amounts in analogous reactions run in homogeneous phase. Additional interest arises in chiral catalysts bonded to polymeric matrices where the catalyst is recycled for an indefinite number of reactions. This means that even very sophisticated and expensive catalysts could be used for large-scale preparations, while avoiding tedious operations of separation of the products from the optically active catalysts.

Only a limited number of examples of the application of PTC to asymmetric synthesis have been reported.

The first is the alkylation of oxo esters with allyl bromide in the presence of catalytic amounts of (—)N-benzyl-N-methylephedrinium bromide *40a* [281]. The enantiomeric excess in the adduct has been estimated to be 5 to 6%.

The reduction of prochiral and/or sterically hindered ketones with BH_4^- and ammonium cations *40a, b* derived from ephedrine afforded optically active alcohols in optical yields up to 13.7% [94, 282]; the addition of nitromethane to chalcone in the presence of KF occurs with an enantiomeric excess up to 26.2% [283, 284]. In the case of the reduction of ketones, the asymmetric induction dropped to values very close to zero for substrates lacking appreciable steric hindrance near the carbonyl group such as 2-octanone, acetophenone or propiophenone [93] (Eq. (30)).

$$R'-\underset{\underset{O}{\|}}{C}-R'' + NaBH_4 \xrightarrow[\text{benzene}/H_2O]{40a, b} R'-\underset{\underset{OH}{|}}{\overset{*}{C}H}-R'' \qquad (30)$$

Starting from the same substrates and similar catalysts, other authors claimed remarkably higher optical yields (up to 39% of enantiomeric excess), particularly by working with larger amounts of catalysts [285]; these results, however, could not be reproduced [282].

$$Ph-\underset{\underset{R}{|}}{CH}-CH(CH_3)-N^+(CH_3)_2R'\ X^- \qquad Ph-\underset{\underset{OH}{|}}{CH}-CH_2CH_2-N^+(CH_3)_2C_{12}H_{25}\ X^-$$

40a, R=OH, R'=PhCH$_2$
40b, R=OH, R'=C$_{12}$H$_{25}$
40c, R=H, R'=C$_{12}$H$_{25}$

41

$$\underset{Ph}{H}\diagdown\underset{O}{\triangle}\diagup\underset{H}{CH_3}$$

42

The presence of the OH group β to the quaternary nitrogen was found to be essential for achieving an asymmetric synthesis. Onium salts *41* with a hydroxy group γ to the quaternary group gave very low optical yields; catalysts without OH (*40c*) gave no optical yield. Interestingly, a hydroxy group in the β-position also increases the reaction rate [94] (see Sect. 7.9.1).

In basic medium, ephedrinium salts decompose to oxiranes *42* which exhibit a very high rotatory power and, even if present in small amounts in the products, are a source of error in evaluations of asymmetric synthesis [286, 287].

Quibec *43a*, introduced by Wynberg [288], proved to be the most effective catalyst for asymmetric synthesis in PTC. In particular, in the epoxidation of the chalcone (Eq. (31)) with alkaline 30% H_2O_2 in the presence of catalytic amounts of *43a*,

43a, R=H;   *43b*, R=NO_2

optical yields up to 54% were obtained [289]. A mechanism for this reaction has been proposed [290].

$$o\text{-}CH_3O\text{-}C_6H_4\text{-}CO\text{-}CH=CH\text{-}Ph \xrightarrow[\text{43a}]{HO_2^-,\ toluene/H_2O} o\text{-}CH_3O\text{-}C_6H_4\text{-}CO\text{-}CH(O)CH\text{-}Ph \quad (31)$$
54% e.e.

In the reduction of ketones with BH_4^- and *43a* as catalyst, phenyl-*t*-butylcarbinol was obtained with a 32% e.e. [282].

The corresponding *o*-nitrobenzyl chloride *43b* catalyzed the asymmetric addition of thiophenol to 2-cyclohexen-1-one under solid-liquid PTC conditions (e.e. up to 35.6%) [284].

Darzens' reactions of aromatic aldehydes and α-chloroacetonitriles, α-chloroalkyl sulfones, or phenacyl halides with ephedrine or quinine derivatives as catalysts afforded epoxides in low optical yields [291-293].

Chiral catalysts bonded to polymeric matrices have also been used in asymmetric synthesis. Phenylacetonitrile has been alkylated with such catalysts; however, optical yields were very low [294].

In a Darzens' condensation the optical yield increased up to 23% in the presence of chiral catalysts bonded to a polymer. This yield is much higher than that obtained with analogous soluble catalysts [291]. In other cases, the progressive increase of the spacer chain length resulted in no optical yield [295].

9 References

1. Parker, A. J.: Chem. Rev. *69*, 1 (1969)
2. Pedersen, C. J.: J. Am. Chem. Soc. *89*, 2495, 7017 (1967)
3. Pedersen, C. J., Frensdorff, H. K.: Angew. Chem., Int. Ed. Engl. *11*, 16 (1972)
4. Dietrich, B., Lehn, J.-M., Sauvage, J. P.: Tetrahedron Lett. *1969*, 2885, 2889
5. Lehn, J.-M.: Structure and Bonding *16*, 1 (1973)
6. Makosza, M., Serafin, B.: Rocz. Chem. *39*, 1223 (1965)
7. Makosza, M.: Pure Appl. Chem. *43*, 439 (1975)
8. Brändström, A., Gustavii, K.: Acta Chem. Scand. *23*, 1215 (1969)
9. Brändström, A.: Preparative Ion-Pair Extraction. Läkemedel: Apotekarsocieteten, AB Hässle 1974
10. Starks, C. M., Napier, D. R.: Ital. Patent 832, 967 (1968), Brit. Patent 1, 227, 144 (1971), French Patent 1, 573, 164 (1969); Chem. Abstr. *72*, 115271 (1970)

11. Starks, C. M.: J. Am. Chem. Soc. *93*, 195 (1971)
12. Starks, C. M., Owens, R. M.: J. Am. Chem. Soc. *95*, 3613 (1973)
13. Jarousse, J.: C.R. Acad. Sci. Paris, Ser. C *232*, 1424 (1951)
14. Germ. Offenl. 268, 621, to BASF (1913); Chem. Zentr. *1914*, 310
15. Gibson, N. A., Hosking, J. W.: Aust. J. Chem. *18*, 123 (1965)
16. Dockx, J.: Synthesis *1973*, 441
17. Dehmlow, E. V.: Angew. Chem., Int. Ed. Engl. *13*, 170 (1974)
18. Montanari, F.: Chim. Ind. (Milan) *57*, 17 (1975)
19. Makosza, M.: Naked anions-phase transfer. In: Modern Synthetic Methods 1976. Scheffold, R. (ed.), pp. 7–100. Zürich: Schweizerischer Chemiker-Verband 1976
20. Dou, H. J.-M.: Chimie Actualités *1976*, 41
21. Jones, R. A.: Aldrichimica Acta *9*, 35 (1976)
22. Schacht, E.: Kontakte *1976*, 3
23. Dehmlow, E. V.: Angew. Chem., Int. Ed. Engl. *16*, 493 (1977)
24. Makosza, M.: Russ. Chem. Rev. *46*, 1151 (1977)
25. Brändström, A.: Principles of phase-transfer catalysis by quaternary ammonium salts. In: Adv. Phys. Org. Chem., vol. 15, Fold, V. (ed.), pp. 267–330. New York: Academic Press 1977
26. Varughese, P.: J. Chem. Educ. *54*, 666 (1977)
27. Jones, R. C. F.: Gen. Synth. Methods *1*, 424 (1978)
28. McIntosh, J. M.: J. Chem. Educ. *55*, 235 (1978)
29. Weber, W. P., Gokel, G. W.: J. Chem. Educ. *55*, 350, 429 (1978)
30. Dou, H. J.-M.: Actual Chim. *1978*, 7
31. Makosza, M.: Two-phase reactions in organic chemistry. In: Survey of Progress in Chemistry, vol. 9. New York: Academic Press 1979
32. Keller, V. E.: Compendium of phase-transfer reactions and related synthetic methods. Buchs: Fluka AG 1979
33. Dehmlow, E. V.: Chimia *34*, 12 (1980)
34. D'Incan, E., Viout, P.: Techniques Ing. *J-1192*, 1 (1980)
35. Weber, W. P., Gokel, G. W.: Phase transfer catalysis in organic synthesis, Berlin-Heidelberg-New York: Springer-Verlag 1977
36. Starks, C. M., Liotta, C.: Phase transfer catalysis. Principles and techniques. New York: Academic Press 1978
37. Dehmlow, E. V., Dehmlow, S. S.: Phase transfer catalysis. Weinheim: Verlag Chemie 1980
38. Gokel, G. W., Durst, H. D.: Synthesis *1976*, 168
39. Knipe, A. C.: J. Chem. Educ. *53*, 618 (1976)
40. Izatt, R. M., Christensen, J. J.: Synthetic multidentate macrocyclic compounds. New York: Academic Press 1978
41. Lehn, J.-M.: Acc. Chem. Res. *11*, 49 (1978)
42. Lehn, J.-M.: Pure Appl. Chem. *52*, 2303 (1980)
43. Bradshaw, J. S., Stott, P. E.: Tetrahedron *36*, 461 (1980)
44. Weber, E., Vögtle, F.: Kontakte *1977*, 26
45. Weber, E., Vögtle, F.: Kontakte *1977*, 48
46. Weber, E., Vögtle, F.: Kontakte *1978*, 16
47. Vögtle, F., Weber, E., Elben, U.: Kontakte *1980*, 36
48. Caubère, P.: Acc. Chem. Res. *7*, 301 (1974)
49. Meyers, C. Y. et al.: New syntheses and reactions of organic compounds: reactions with carbon tetrachloride and other perhalomethanes in powdered potassium hydroxide-t-butyl alcohol. In: Catalysis in Organic Synthesis, pp. 197–278. New York: Academic Press 1977
50. Meyers, C. Y.: New reactions and syntheses of organic compounds. Reactions of sulfones with CCl_4 and other perhalomethanes in KOH-t-BuOH. In: Topics in Organic Sulfur Chemistry. Tishler, M. (ed.), pp. 207–260. Ljubljana: University Press 1978
51. Modin, R., Schill, G.: Acta Pharm. Suec. *4*, 301 (1967)
52. Modin, R.: Acta Pharm. Suec. *9*, 157 (1972), and references therein
53. Gustavii, K.: Acta Pharm. Suec. *4*, 233 (1967)
54. Gustavii, K., Schill, G.: Acta Pharm. Suec. *3*, 241 (1966)
55. Ref. [37], p. 14

56. Gordon, J. E., Kutina, R. E.: J. Am. Chem. Soc. *99*, 3903 (1977)
57. Dehmlow, E. V., Slopianka, M., Heider, J.: Tetrahedron Lett. *1977*, 2361
58. Kheifets, V. L., Yakovleva, N. A., Krasil'schik, B. Ya.: Zh. Prikl. Khim. *46*, 549 (1973); Chem. Abstr. *79*, 10505 (1973)
59. Landini, D., Maia, A., Montanari, F.: J. Am. Chem. Soc. *100*, 2796 (1978)
60. Landini, D., Maia, A., Montanari, F.: Nouv. J. Chim. *3*, 575 (1979)
61. Landini, D., Maia, A., Montanari, F.: unpublished results
62. Makosza, M., Fedorynski, M.: Synth. Commun. *3*, 305 (1973)
63. Agarwal, B. R., Diamond, R. M.: J. Phys. Chem. *67*, 2785 (1963)
74. Freedman, H. H., Dubois, R. A.: Tetrahedron Lett. *1975*, 3251
65. Antoine, J. P. et al.: Bull. Soc. Chim. Fr. *1980-II*, 207
66. Herriott, A. W., Picker, D.: J. Am. Chem. Soc. *97*, 2345 (1975)
67. Landini, D., Maia, A., Montanari, F.: J. Chem. Soc., Chem. Commun. *1977*, 112
68. Landini, D. et al.: J. Chem. Soc., Chem. Commun. *1975*, 950
69. Menger, F. N.: J. Am. Chem. Soc. *92*, 5965 (1970)
70. Fendler, J. H., Fendler, E. J.: Catalysis in micellar and macromolecular systems. New York: Academic Press 1975
71. Horner, L., Gerhard, J.: Justus Liebigs Ann. Chem. *1980*, 838
72. Ref. [36], p. 32
73. Uglestad, J., Ellingsen, T., Berge, A.: Acta Chem. Scand. *20*, 1592 (1966)
74. Bhattacharyya, D. N. et al.: J. Phys. Chem. *69*, 608 (1965)
75. Kraus, C. A.: J. Phys. Chem. *60*, 129 (1956)
76. Davies, M., Williams, G.: Trans. Faraday Soc. *56*, 1619 (1960)
77. Gustavii, K., Schill, G.: Acta Pharm. Suec. *3*, 259 (1966)
78. Guibe, F., Bram, G.: Bull. Soc. Chim. Fr. *1975*, 933
79. Illuminati, G.: Solvent effects on selected organic and organometallic reactions. Guidelines to synthetic applications. In: Chemistry 8/2, Dack M. R. J. (ed.), pp. 159–233. New York: Wiley 1976
80. Brändström, A., Kalind-Andersen, H.: Acta Chem. Scand. Ser. B *29*, 201 (1975)
81. Brändström, A.: Acta Chem. Scand., Ser. B. *30*, 203 (1976)
82. Zanger, M., Van der Werf, C. A., McEwen, W. E.: J. Am. Chem. Soc. *81*, 3806 (1959)
83. McEwen, W. E. et al.: J. Am. Chem. Soc. *86*, 2378 (1964)
84. McEwen, W. E. et al.: J. Am. Chem. Soc. *87*, 3948 (1965)
85. Pagilagan, R. U., McEwen, W. E.: Chem. Commun. *1966*, 652
86. Campbell, J. B.: U.S. Patent 3, 639, 492 (1972); Chem. Abstr. *77*, 21 156 (1972)
87. Pocker, Y., Parker, A. J.: J. Org. Chem. *31*, 1526 (1966)
88. Islam, Md. N., Leffek, K. T.: J. Chem. Soc., Perkin Trans. 2 *1977*, 958
89. Landini, D., Maia, A., Montanari, F.: Nouv. J. Chim. *4*, 723 (1980)
90. Yano, Y., Okonogi, T., Tagaki, W.: J. Org. Chem. *38*, 3912 (1973)
91. Umezawa, T., Okonogi, T., Tagaki, W.: J. Chem. Soc., Chem. Commun. *1974*, 363
92. Yano, Y., Okonogi, T., Tagaki, W.: Bull. Chem. Soc. Japan *47*, 771 (1974)
93. Colonna, S., Fornasier, R.: Synthesis *1975*, 531
94. Balcells, J., Colonna, S., Fornasier, R.: Synthesis *1976*, 266
95. Ref. [37], p. 28
96. Makosza, M., Bialecka, E.: Tetrahedron Lett. *1977*, 183
97. Dehmlow, E. V., Lissel, M.: Tetrahedron Lett. *1976*, 1783
98. Dehmow, E. V., Lissel, M., Heider, J.: Tetrahedron *33*, 363 (1977)
99. Brändström, A., Junggren, U., Lamm, B.: Tetrahedron Lett. *1972*, 3173
100. Dehmlow, E. V.: Tetrahedron Lett. *1976*, 91
101. Dehmlow, E. V., Remmler, T.: J. Chem. Res. *1977*, (S) 72, (M) 766
102. Landini, D., Rolla, F.: Synthesis *1976*, 389
103. Fedorynsky, M. et al.: J. Org. Chem. *43*, 4682 (1978)
104. Liotta, C. L., Harris, H. P.: J. Am. Chem. Soc. *96*, 2250 (1974)
105. Cook, F. L., Bowers, C. W., Liotta, C. L.: J. Org. Chem. *39*, 3416 (1974)
106. Tundo, P.: J. Org. Chem. *44*, 2048 (1979)
107. Tundo, P., Venturello, P.: Synthesis *1979*, 952
108. Angeletti, E., Tundo, P., Venturello, P.: J. Chem. Soc., Chem. Commun. *1980*, 1127

109. Angeletti, E., Tundo, P., Venturello, P.: private communication
110. Landini, D. et al.: J. Chem. Soc., Perkin Trans. 2 *1980*, 46
111. Stott, P. E., Bradshaw, J. S., Parish, W. W.: J. Am. Chem. Soc. *102*, 4810 (1980)
112. Landini, D., Montanari, F., Pirisi, F. M.: J. Chem. Soc., Chem. Commun. *1974*, 879
113. Landini, D. et al.: Gazz. Chim. Ital. *105*, 863 (1975)
114. Makosza, M., Ludwikow, M.: Angew. Chem., Int. Ed. Engl. *13*, 665 (1974)
115. Cinquini, M., Tundo, P.: Synthesis *1976*, 516
116. Montanari, F., Tundo, P.: Tetrahedron Lett. *1979*, 5055
117. Wong, K. H.: J. Chem. Soc., Chem. Commun. *1978*, 282
118. Mizuno, T. et al.: Bull. Chem. Soc. Japan *53*, 481 (1980)
119. Mathias, L. J., Burkett, D.: Tetrahedron Lett. *1979*, 4709
120. Yamashita, J., Ishikawa, S., Hashimoto, H.: Bull. Chem. Soc. Japan *53*, 736 (1980)
121. Ref. [37], p. 20
122. Vander Zwan, M. C., Hartner, F. W.: J. Org. Chem. *43*, 2655 (1978)
123. Dietrich, B., Lehn, J.-M.: Tetrahedron Lett. *1973*, 1225
124. Cinquini, M., Montanari, F., Tundo, P.: J. Chem. Soc., Chem. Commun. *1975*, 393
125. Akabori, S., Ohtomi, M.: Bull. Chem. Soc. Japan *48*, 2991 (1975)
126. Jeanne, F., Trichet, A.: Compte Rendue de Fin d'Etude, Action Concertée: Activation Selective en Chimie Organique (Catalyse Homogène) 1975
127. Clement, D., Damm, F., Lehn, J.-M.: Heterocycles *5*, 477 (1976)
128. Cinquini, M., Montanari, F., Tundo, P.: Gazz. Chim. Ital. *107*, 11 (1977)
129. Akabori, S., Tuji, H.: Bull. Chem. Soc. Japan *51*, 1197 (1978)
130. Landini, D., Montanari, F., Rolla, F.: Synthesis *1978*, 223
131. Buhleier, E., Wehner, W., Vögtle, F.: Chem. Ber. *112*, 546 (1979)
132. Landini, D. et al.: J. Am. Chem. Soc. *101*, 2526 (1979)
133. Taft, R. W.: Proton transfer equilibria in the gas and solution phases. In: Kinetics of Ion-Molecule Reactions, NATO advanced study, Institute series (B-physics). Ausloss, P. (ed.) *40*, 271 (1979)
134. Landini, D. et al.: J. Chem. Soc., Perkin Trans. 2 *1981*, 821
135. For this nomenclature, see Ref. [136]
136. Vögtle, F., Weber, E.: Angew. Chem., Int. Ed. Engl. *18*, 753 (1979)
137. Menger, F. M., Rhee, J. U., Rhee, H. K.: J. Org. Chem. *40*, 3803 (1975)
138. Lee, D. G., Chang, V. S.: J. Org. Chem. *43*, 1532 (1978)
139. Sanger, W., Suh, I.-H., Weber, G.: Isr. J. Chem. *18*, 253 (1979)
140. Vögtle, F., Heimann, U.: Chem. Ber. *111*, 2757 (1978)
141. Lehmkuhl, H., Rabet, F., Hauschild, K.: Synthesis *1977*, 184
142. Yanagida, S., Noji, Y., Okahara, M.: Tetrahedron Lett. *1977*, 2893
143. Aageev, F. Kh. et al.: Azerb. Khim. Zh. *1977*, 39; Chem. Abstr. *87*, 117514 (1978)
144. Movsumzade, M. M. et al.: Zh. Org. Khim. *1976*, 2477; Chem. Abstr. *86*, 71795 (1977)
145. Hirao, A. et al.: Makromol. Chem. *179*, 915 (1978)
146. Hirao, A. et al.: Makromol. Chem. *179*, 1735, 2343 (1978)
147. Töke, L., Szabo, G. T., Somogy-Warner, K.: Acta Chim. Sci. Hung. *101*, 47 (1979); Chem. Abstr. *92*, 65502 (1980)
148. Balasubramanian, D., Sukumar, P., Chandani, B.: Tetrahedron Lett. *1979*, 3543
149. Kitazume, T., Ishikawa, N.: Chem. Lett. *1978*, 283
150. Bartsch, R. A., Yang, I. W.: Tetrahedron Lett. *1979*, 2503
151. Vögtle, F., Weber, E.: Angew. Chem., Int. Ed. Engl. *13*, 814 (1974)
152. Knöchel, A., Oehler, J., Rudolph, G.: Tetrahedron Lett. *1975*, 3167
153. Akabori, S., Ohtomi, M., Yatabe, S.: Bull. Chem. Soc. Japan *53*, 1463 (1980)
154. Hyatt, J. A.: J. Org. Chem. *43*, 1808 (1978)
155. Frensch, K., Vögtle, F.: Liebigs Ann. Chem. *1979*, 2121
156. Fornasier, R. et al.: Tetrahedron Lett. *1976*, 1381
157. Normant, H., Cuvigny, T., Savignac, P.: Synthesis *1975*, 805
158. Tomoi, M. et al.: Chem. Lett. *1976*, 473
159. Tomoi, M. et al.: Bull. Chem. Soc. Japan *52*, 1653 (1979)
160. Mikolajczyk, M. et al.: Tetrahedron Lett. *1975*, 3757
161. Samaan, S., Rolla, F.: Phosphorus Sulfur *4*, 145 (1978)

162. Gokel, G. W., Garcia, B. J.: Tetrahedron Lett. *1978*, 1743
163. Hayashi, Y.: Japan Kokai 77/111, 486; Chem. Abstr. *88*, 104653 (1978)
164. Garcia, B. J., Leopold, A., Gokel, G. W.: Tetrahedron Lett. *1980*, 2115
165. Tanaka, T., Mukaiyama, T.: Chem. Lett. *1976*, 1259
166. Armstrong, D. W., Kornahrens, H.: Tetrahedron Lett. *1979*, 4525
167. Dietrich, B. et al.: Helv. Chim. Acta *62*, 2763 (1979)
168. Manecke, G., Storck, W.: Angew. Chem., Int. Ed. Engl. *17*, 657 (1978)
169. Regen, S. L.: Angew. Chem., Int. Ed. Engl. *18*, 421 (1979)
170. Regen, S. L.: J. Am. Chem. Soc. *97*, 5956 (1975)
171. Regen, S. L.: J. Org. Chem. *42*, 875 (1977)
172. Zadeh, H. K., Dou, H. J.-M., Metzger, J.: J. Org. Chem. *43*, 156 (1978)
173. Cinquini, M. et al.: J. Chem. Soc., Chem. Commun. *1976*, 394
174. Molinari, H. et al.: J. Am. Chem. Soc. *101*, 3920 (1979)
175. Brown, J. M., Jenkins, J. A.: J. Chem. Soc., Chem. Commun. *1976*, 458
176. Molinari, H., Montanari, F., Tundo, P.: J. Chem. Soc., Chem. Commun. *1977*, 639
177. Chiles, M. S., Jackson, D. D., Reeves, P. C.: J. Org. Chem. *45*, 2915 (1980)
178. Sparrow, J. T.: J. Org. Chem. *41*, 1350 (1976)
179. Blasius, E. et al.: Z. Anal. Chem. *284*, 337 (1977)
180. Blasius, E., Janzen, K. P., Neumann, W.: Mikrochim. Acta *2*, 279 (1977)
181. Blasius, E., Maurer, P. G.: Makromol. Chem. *178*, 649 (1977)
182. Takaki, V., Smid, J.: J. Am. Chem. Soc. *96*, 2588 (1974)
183. Kopolow, S., Hogen Esch, T. E., Smid, J.: Macromolecules *6*, 133 (1973)
184. See Ref. [43], p. 504
185. Montanari, F., Tundo, P.: J. Org. Chem. *46*, 2125 (1981)
186. Tomoi, M. et al.: Tetrahedron Lett. *1978*, 3031
187. Tomoi, M., Kihara, K., Kakiuchi, H.: Tetrahedron Lett. *1979*, 3485
188. Smid, J.: Pure Appl. Chem. *48*, 343 (1976)
189. Bogatskii, A. V., Lukyanenko, N. G., Pastushok, V. N.: Dokl. Akad. Nauk SSSR *247*, 1153 (1979); Chem. Abstr. *92*, 6206 (1980)
190. Regen, S. L., Dulak, L.: J. Am. Chem. Soc. *99*, 623 (1977)
191. Regen, S. L., Besse, J. J., McLick, J.: J. Am. Chem. Soc. *101*, 116 (1979)
192. Hiratani, K., Reuter, P., Manecke, G.: Isr. J. Chem. *18*, 208 (1979)
193. Regen, S. L.: J. Am. Chem. Soc. *99*, 3838 (1977)
194. Regen, S. L., Nigam, A., Besse, J. J.: Tetrahedron Lett. *1978*, 2757
195. Maeda, H., Hayashi, Y., Teramura, K.: Chem. Lett. *1980*, 677
196. McKenzie, W. M., Sherrington, D. C.: J. Chem. Soc., Chem. Commun. *1978*, 541
197. Yanagida, S., Takahashi, K., Okahara, M.: J. Org. Chem. *44*, 1099 (1979)
198. Montanari, F., Quici, S.: unpublished results
199. Regen, S. L., Besse, J. J.: J. Am. Chem. Soc. *101*, 4059 (1979)
200. Tundo, P.: Synthesis *1978*, 315
201. D'Incan, E., Viout, P.: Tetrahedron *31*, 159 (1975)
202. Kornblum, N., Seltzer, R., Haberfield, P.: J. Am. Chem. Soc. *85*, 1148 (1963)
203. Rolla, F., Roth, W., Horner, L.: Naturwissenschaften *64*, 377 (1977)
204. Tundo, P.: J. Chem. Soc., Chem. Commun. *1977*, 641
205. Tundo, P., Venturello, P.: J. Am. Chem. Soc. *101*, 6606 (1979)
206. Tundo, P., Venturello, P.: J. Am. Chem. Soc. *103*, 856 (1981)
207. Dou, H. J.-M. et al.: J. Org. Chem. *42*, 4275 (1977)
208. Landini, D., Rolla, F.: Synthesis *1974*, 565
209. Landini, D., Rolla, F.: Synthesis *1976*, 389
210. Landini, D., Quici, S., Rolla, F.: Synthesis *1975*, 430
211. Corey, E. J. et al.: Tetrahedron Lett. *1975*, 3183
212. San Filippo, J., Chern, C., Valentino, J. S.: J. Org. Chem. *40*, 1678 (1975)
213. Johnson, R. A., Nidy, E. G.: J. Org. Chem. *40*, 1680 (1975)
214. See, *inter alia* Refs. [32]; [35], pp. 73, 85, 96, 109, 117; [36], p. 91, [37], p. 56
215. Makosza, M.: Tetrahedron Lett. *1969*, 673
216. Makosza, M. et al.: Tetrahedron *30*, 3723 (1974)
217. Gokel, G. W., Korzeniowski, S. H., Blum, L.: Tetrahedron Lett. *1977*, 1633

218. Korzeniowski, S. H., Gokel, G. W.: Tetrahedron Lett. *1977*, 1637
219. Markezich, R. L. et al.: J. Org. Chem. *42*, 3435 (1977)
220. Quan, P. M., Korn, S. R.: Germ. Offenl. 2, 634, 419 (1977); Chem. Abstr. *87*, 52942 (1977)
221. Reeves, W. P., Simmons, A., Keller, W.: Synth. Commun. *10*, 633 (1980)
222. Montanari, F. et al.: unpublished results
223. Wilczynski, W., Jawdosiuk, M., Makosza, M.: Rocz. Chem. *51*, 1643 (1977)
224. Jawdosiuk, M. et al.: Pol. J. Chem. *52*, 2189 (1978)
225. Serio Duggan, A. J., Grabowski, E. J. J., Russ, W. K.: Synthesis *1980*, 573
226. Maiorana, S. et al.: unpublished results
227. Sam, D. J., Simmons, H. E.: J. Am. Chem. Soc. *96*, 2252 (1974)
228. Jackman, L. M., Lange, B. C.: Tetrahedron *33*, 2737 (1977)
229. Bram, G.: J. Molecular Cat., in press
230. See, *inter alia*, Refs. [32]; [35], p. 136; [36], p. 170; [37], p. 101
231. See, *inter alia*, Refs. [32]; [35], p. 125; [36], p. 330; [37], p. 149
232. Landini, D., Quici, S., Rolla, F.: Synthesis *1975*, 397
233. Naso, F., Ronzini, L.: J. Chem. Soc., Perkin Trans. 1 *1974*, 340
234. Chollet, A., Hagenbuch, J. P., Vogel, P.: Helv. Chim. Acta *62*, 511 (1979)
235. Makosza, M., Kacprowicz, A., Fedorynski, M.: Tetrahedron Lett. *1975*, 2119
236. See, *inter alia*, Refs. [32]; [35], pp. 18, 58; [36], p. 224; [37], pp. 177, 239
237. See, *inter alia*, Refs. [32]; [35], pp. 134, 234, 252; [36], pp. 288, 341; [37], p. 133
238. Landini, D., Montanari, F., Rolla, F.: Synthesis *1974*, 37
239. Landini, D., Montanari, F., Rolla, F.: Synthesis *1978*, 771
240. Landini, D., Rolla, F.: J. Org. Chem. *45*, 3527 (1980)
241. Rolla, F., Landini, D., Montanari, F.: Inorg. Chim. Acta *40*, X138 (1980)
242. Dehmlow, E. V., Slopianka, M.: Chem. Ber. *112*, 2765 (1979)
243. Cousseau, J.: Synthesis *1980*, 805
244. Sam, D. J., Simmons, H. E.: J. Am. Chem. Soc. *94*, 4024 (1972)
245. Weber, W. P., Shepherd, J. P.: Tetrahedron Lett. *1972*, 4907
246. Okimoto, T., Swern, D.: J. Am. Oil Chem. Soc. *54*, 867 A (1977)
247. Dou, H. J.-M., Komeili-Zadeh, H., Crozet, C.: C. R. Acad. Sci. Paris, Ser. C. *284*, 685 (1977)
248. Pletcher, D., Tait, S. J. D.: J. Chem. Soc., Perkin Trans. 2 *1979*, 788
249. Landini, D., Montanari, F., Rolla, F.: Synthesis *1979*, 134
250. Landini, D., Rolla, F.: Chem. Ind. (London) *1980*, 213
251. Lee, G. A., Freedman, H. H.: Tetrahedron Lett. *1976*, 1641
252. Sy, A. D., Raksis, J. W.: Tetrahedron Lett. *1980*, 2223
253. See, *inter alia*, Refs. [32]; [35], pp. 109, 206; [36], p. 298; [37], p. 249
254. Rylander, P. N.: Organic synthesis with noble metal catalysts, pp. 121–144. New York: Academic Press 1973
255. Lee, D. G., Van den Engh, M.: The oxidation of organic compounds by ruthenium tetroxide. In: Oxidation in Organic Chemistry. Trahanovsky, W. S. (ed.), pp. 177–227. New York: Academic Press 1973
256. Starks, C. M., Napier, D. R.: S. African Patent 7, 101, 495 (1971), Brit. Patent 1, 324, 763 (1973); Chem. Abstr. *76*, 153191 (1972)
257. Starks, C. M., Washecheck, P. H.: U.S. Patent 3,547,962 (1970); Chem. Abstr. *74*, 140895 (1971)
258. Herriott, A. W., Picker, D.: Tetrahedron Lett. *1974*, 1511
259. Napier, D. R., Starks, C. M.: U.S. Patent 3, 992, 432 (1976); Chem. Abstr. *86*, 34757 (1977)
260. Ho, T.-L., Gupta, B. G. B., Olah, G. A.: Synthesis *1977*, 676
261. Alneri, E., Bottaccio, G., Carletti, V.: Tetrahedron Lett. *1977*, 2117
262. Sudnes, L. K.: Acta Chem. Scand., Ser. B. *31*, 903 (1977)
263. Pierre, J. L., Handel, H.: Tetrahedron Lett. *1974*, 2317
264. Pierre, J. L., Handel, H., Perraud, R.: Tetrahedron *31*, 2795 (1975)
265. Handel, H., Pierre, J. L.: Tetrahedron *31*, 2799 (1975)
266. Loupy, A., Seyden-Penne, J., Tchoubar, B.: Tetrahedron Lett. *1976*, 1677
267. Brändström, A., Jumgren, H., Lamm, B.: Tetrahedron Lett. *1972*, 3173
268. Rolla, F.: unpublished results
269. Borgogno, G., Colonna, S., Fornasier, R.: Synthesis *1975*, 529

270. Des Abbayes, H., Alper, H.: J. Am. Chem. Soc. *99*, 98 (1977)
271. Reger, D. L., Habib, M. M., Fauth, D. J.: Tetrahedron Lett. *1979*, 115
272. Cassar, L.: Ann. N.Y. Acad. Sci. *333*, 208 (1980)
273. Cassar, L., Foà, M., Giordano, A.: J. Organomet. Chem. *121*, C 55 (1976)
274. Cassar, L., Foà, M.: J. Organomet. Chem. *134*, C 15 (1977)
275. Alper, H., Des Abbayes, H.: J. Organomet. Chem. *134*, C 11 (1977)
276. Des Abbayes, H., Buloup, A.: J. Chem. Soc., Chem. Commun. *1978*, 1090
277. Foà, M., Cassar, L.: Gazz. Chim. Ital. *109*, 619 (1979)
278. Cassar, L. et al.: J. Organomet. Chem. *173*, C 335 (1979)
279. Yamamura, K., Murakashi, S. I.: Tetrahedron Lett. *1977*, 4429
280. Salisova, M., Alper, H.: Angew. Chem., Int. Ed. Engl. *18*, 792 (1979)
281. Fiaud, J.-C.: Tetrahedron Lett. *1975*, 3495
282. Colonna, S., Fornasier, R.: J. Chem. Soc., Perkin Trans. 1 *1978*, 371
283. Colonna, S., Hiemstra, H., Wynberg, H.: J. Chem. Soc., Chem. Commun. *1978*, 238
284. Colonna, S., Re, A., Wynberg, H.: J. Chem. Soc., Perkin Trans. 1 *1981*, 547
285. Massé, J. P., Parayre, E. R.: J. Chem. Soc., Chem. Commun. *1976*, 438
286. Hiyama, T. et al.: J. Am. Chem. Soc. *98*, 1641 (1976)
287. Hiyama, T. et al.: J. Am. Chem. Soc. *97*, 1626 (1975)
288. Helder, R. et al.: Tetrahedron Lett. *1976*, 1831
289. Wynberg, H., Greijdanus, B.: J. Chem. Soc., Chem. Commun. *1978*, 427
290. Ref. [37], p. 256
291. Colonna, S., Fornasier, R., Pfeiffer, U.: J. Chem. Soc., Perkin Trans. 1 *1978*, 8
292. Hummelen, J. C., Wynberg, H.: Tetrahedron Lett. *1978*, 1089
293. Annunziata, R.: Synth. Commun. *9*, 171 (1979)
294. Chiellini, E., Solaro, R.: J. Chem. Soc., Chem. Commun. *1977*, 231
295. Colonna, S.: private communication

Closing Remarks

Together with the introductory overview in volume 98, the more specific chapters are intended to report recent results and research trends and to install a link to other fields less fully represented in these two volumes.

The design of receptor cavities for many convex molecules and anions is conceivable. A particularly important application of concave cavities would be those which could very selectively take up urea, glucose, creatinine, specific peptides and other physiologically relevant medio molecules, drug metabolites, etc. Bound to resins, such receptors could be useful in removing urea from blood circulation and thus in promoting the dialysis function of the kidney. Similarly, selective glucose receptors could lower blood-sugar levels. Other potential applications are the control of the transport of neutral particles like sugar, adrenalin, noradrenalin, etc. across membranes with the help of suitable receptor cavities; and the steering of anion transport by anion carriers.

It is fascinating to visualize the complex chemistry of new macro-cavities, which, by virtue of their almost infinite variability, are able to take up bigger convex molecules, organic cations and anions. There is hardly any doubt that the mostly convex-oriented organic chemistry of today ("convex chemistry") will be complemented with an opposite chemistry of concave receptor molecules ("concave chemistry").

Presumably, this will lead to the ability to stabilize a series of less stable inorganic and organic cationic, anionic and neutral convex compounds by molecular encapsulation. We can foresee phase transfer reactions and phase transfer catalysis with masked uncharged organic molecules modified in this way in their solubility and reactivity. Approved reagents may be modified with regard to reactivity and selectivity of the reaction.

Reactions involving one or more reactants enclosed separately or together, and in this way protected in big tailor-shaped cavities (resembling the "Mobil process" using zeolite cavities in the production of petrol from methanol) with varied kinetics, stereochemistry, and selectivity, are conceivable. Thus, new types of selective enzyme-analoguous reactions may be studied and used.

Whom may we credit for the opening of this field, with so many potential research aspects? The innovation emerged from the discovery of the crown ethers as well as from progress in synthesis of medium and large rings (where high yields have become routine), progress in enzyme model reactions and in membrane research, molecular model development, and physical-organic — including spectroscopic methods. In this field we shall be observing the progress of an interesting new branch in chemistry.

F. Vögtle/Bonn, FRG

Author Index Volume 101

Contents of Vols. 50–100 see Vol. 100
Author and Subject Index Vols. 26–50 see Vol. 50

The volume numbers are printed in italics

Hilgenfeld, R., and Saenger, W.: Structural Chemistry of Natural and Synthetic Ionophores and their Complexes with Cations. *101*, 3–82 (1982).

Kellogg, R. M.: Bioorganic Modelling — Stereoselective Reactions with Chiral Neutral Ligand Complexes as Model Systems for Enzyme Catalysis. *101*, 111–145 (1982).

Landini, D., see Montanari, F.: *101*, 111–145 (1982).

Montanari, F., Landini, D., and Rolla, F.: Phase-Transfer Catalyzed Reactions. *101*, 149–200 (1982).

Painter, R., and Pressman, B. C.: Dynamics Aspects of Ionophore Mediated Membrane Transport. *101*, 84–110 (1982).

Pressman, B. C., see Painter, R.: *101*, 84–110 (1982).

Rolla, F., see Montanari, F.: *101*, 111–145 (1982).

Saenger, W., see Hilgenfeld, R.: *101*, 3–82 (1982).

New Syntheses with Carbon Monoxide

Editor: J. Falbe
With contributions by H. Bahrmann, B. Cornils, C.D. Frohning, A. Mullen
1980. 118 figures, 127 tables. XIV, 465 pages
(Reactivity and Structure, Volume 11)
ISBN 3-540-09674-4

The book is a report on developments in carbon monoxide chemistry, emphasis being placed on advances made in the seventies. Topics such as hydroformylation, Fischer-Tropsch chemistry, carbonylation, S.N.G., as well as ring closure reactions with carbon monoxide are treated in depth. Current mechanistic theories are critically discussed. Themes are of current interest due to revival of synthesis gas chemistry since the oil crisis in 1973. The book recapitulates only very briefly the period covered by Dr. Falbe's previous book **(Carbon Monoxide in Organic Synthesis,** Springer-Verlag, 1970) focussing on current developments. It gives a detailed review of carbon monoxide chemistry and will help to establish a link between research (industrial and academic) and application.

J. Fabian, H. Hartmann

Light Absorption of Organic Colorants

Theoretical Treatment and Empirical Rules
1980. 76 figures, 48 tables. VIII, 245 pages
(Reactivity and Structure, Volume 12)
ISBN 3-540-09914-X

The relationship between light absorption and electronic structure of organic compounds is a productive area of research for chemists and physicists. Its knowledge is important for identifying colored compounds and for designing new materials for textile dyeing, color photography, photosensitization and laser physics.
The book reviews the light absorption characteristics of a great variety of colored organic compounds of practical interest. Absorption data are listed for representative compounds in each series, paying particular attention to the full range of color change brought about by structural modifications. Color structure relationships are rationalized on the basis of certain fundamental chromophores and charge transfer chromophores.

The International Journal for the Polymer Scientist covering all areas of Polymer Science

Polymer Bulletin

ISSN 0170-0839　　　　　　　　Title No. 289

Editors:
Prof. H.-J. Cantow, Makromolekulare Chemie, Universität Freiburg. Prof. J.P. Kennedy, Dept. of Polymer Science, The University of Akron. Prof. T. Saegusa, Dept. Synthetic Chemistry, Kyoto University.

Editorial Board: H. Batzer, Basel; S. Cesća, San Donato Milanese; K. Dušek, Prague; P.J. Flory, Stanford, CA; J. Furukawa, Tokyo; J.E. McGrath, Blacksburg, VA; H.K. Hall, Jr., Tucson, AZ; M.L. Hallensleben, Hannover; H.H. Kausch, Lausanne; T. Kelen, Budapest; M. Kryszewski, Lódź; A. Ledwith, Liverpool; R.W. Lenz, Amherst, MA; E. Maréchal, Paris; J. Meißner, Zürich; A. Nakajima, Kyoto; G. and S. Henrici Olivé, Research Triangle Park, NC; N.A. Platé, Moscow; C.I. Simionescu, Bucureşti; S. Sivaram, Gujarat; D.H. Solomon, Melbourne; H. Tadokoro, Osaka; M. Takayanagi, Fukuoka; I. Uematsu, Tokyo; O. Vogl, Amherst, MA; C. Wippler, Strasbourg; H. Zahn, Aachen

Editorial Assistant: A. Heinrich, Springer-Verlag Heidelberg

Character: between the purely archival journals of full papers and "letter journals" consisting exclusively of short communications; length of papers, 4–8 pages

High-quality papers with an international spectrum: German-speaking countries, Eastern Europe and Japan 19% each; USA 13%; France 12%; other countries 18%

Competent referee system: rejection rate 35%

Rapid publication of papers: 3–6 weeks

50 reprints of each paper free of charge

No page charge

For subscription information and sample copy write to:
Springer-Verlag, Journal Promotion Dept., P.O. Box 105280, D-6900 Heidelberg, FRG

Springer-Verlag　Berlin　Heidelberg　New York